Rebecca and her rescue dog, Diesel.
Author Photo: Archer Imagery

Rebecca Spyker is a first-time author and full-time corporate anthropologist.

Also occupying her time is her devoted partner, to whom she owes a great deal for his love and support during the writing of this book. She also has three extraordinary rescue dogs, who together weigh one hundred and twenty kilograms.

Bringing joy and calm to her life is a daily mix of Zen meditation, Shinto prayer, Huna healing and yoga, practices and wisdoms she was introduced to during her years in Japan and Hawaii.

Born in England and raised in Australia, Rebecca lived in Japan for an extended period after studying cultural anthropology and Japanese at the University of Western Australia. Captivated and intrigued by Japan's beauty and mystery, it's there she began her fascination with Zen philosophy and Shinto, the country's main religion.

Later, she lived in the tropical paradise of Hawaii, where she studied Huna philosophy and healing.

Rebecca is also a yoga instructor, massage therapist and a hands-on healing practitioner which, these days, she only offers to friends and friends' dogs because she's discovered that writing is much more fun.

She loves to cook and delights in creating delicious vegetarian dishes for friends and family.

In the last decade, Rebecca has become a sought-after business transformation consultant. Drawing on her in-depth knowledge of wisdom traditions helps her effect subtle and positive transformational change.

Art credit to: Lynne Tinley

For the rescued and the rescuers

for whom

love is always the answer

and the answer, always love.

Rebecca Spyker

The Book of Bu-Tails of a Zen Dog

Austin Macauley Publishers
LONDON • CAMBRIDGE • NEW YORK • SHARJAH

Copyright © Rebecca Spyker (2021)

The right of Rebecca Spyker to be identified as author of this work has been asserted by the author in accordance with section 77 and 78 of the Copyright, Designs and Patents Act 1988.

All rights reserved. No part of this publication may be reproduced, stored in a retrieval system or transmitted in any form or by any means, electronic, mechanical, photocopying, recording or otherwise, without the prior permission of the publishers.

Any person who commits any unauthorised act in relation to this publication may be liable to criminal prosecution and civil claims for damages.

All of the events in this memoir are true to the best of author's memory. The views expressed in this memoir are solely those of the author.

A CIP catalogue record for this title is available from the British Library.

ISBN 9781528920124 (Paperback)
ISBN 9781528920131 (Hardback)
ISBN 9781528962995 (ePub e-book)

www.austinmacauley.com

First Published (2021)
Austin Macauley Publishers Ltd
25 Canada Square
Canary Wharf
London
E14 5LQ

I wrote this book to inspire hope and bring love to all who read through the stories of my beloved rescued pooch, Bomber, better known as Bu. This is not a work of fiction; all of Bu's 'tails' are true. They actually happened.

This remarkable dog taught me life lessons that have been as powerful and meaningful as the lessons I've learned from Zen teachers, Shinto priests and Huna shamans.

Bu illustrated how to navigate the funny, sad and perplexing experiences that make up the jigsaw of life. He did this with fortitude, grace and above all, with love. Without his gifts, my life would not be as full.

Most of the adventures Bu and I shared occurred with our friends – Bu's and mine – and I thank them for the care and love shown to us both.

Heartfelt thanks as well to my beloved partner for listening to the book chapters as they evolved, and for spending time in various temples and shalas with me, relearning the profound lessons of patience, compassion and mindfulness.

My champion – you know who you are – I thank you for the encouragement to pursue my dreams and for being the one with the courage that saves, in many ways, in many lands.

A very special thanks to Adam for secret matrices, magical spheres and endlessly enabling flourishing.

A huge thank you to my copy editor Justine for her patience and fortitude.

Finally, thanks to the great team at Austin Macauley for bringing this book to life and Bu back to life once more.

Table of Contents

Book One — 13

Chapter One — 15
Refuge Savage

Chapter Two — 23
Palm Shimmy

Chapter Three — 26
The Poo Connoisseur

Chapter Four — 31
Nannup Tiger

Chapter Five — 36
Bomber's Origins

Chapter Six — 41
Entertain Me Boring Stooge

Chapter Seven — 46
Full Moon Drumming

Chapter Eight — 49
Inorganic Snacks

Chapter Nine 55
The Eighty-Four-Thousand-Dollar Dog

Chapter Ten 60
The Family Jewels

Chapter Eleven 63
For Later On

Chapter Twelve 68
Bribery and Ducks

Chapter Thirteen 72
The Human Whisperer

Chapter Fourteen 78
Jesus Lady!

Chapter Fifteen 83
Yum – First Blood

Chapter Sixteen 88
The Chase

Chapter Seventeen 91
It's My Bed!

Chapter Eighteen 94
The Pack

Chapter Nineteen 100
Sneaking Out

Chapter Twenty 102
The Sky Is Falling!

Chapter Twenty-One 106
Encounters of the Strange Kind

Chapter Twenty-Two 113
Scavenged Prizes

Chapter Twenty-Three 117
By Any Other Name

Chapter Twenty-Four 121
It Wasn't Me!

Chapter Twenty-Five 124
Tin Legs

Chapter Twenty-Six 128
Clean-Up Crew

Chapter Twenty-Seven 131
Analogue Tracking Device

Chapter Twenty-Eight 134
Stairway to Heaven

Chapter Twenty-Nine 137
Resonance Therapy

Chapter Thirty 140
A Few of My Favourite Things

Chapter Thirty-One	**147**
Saying Goodbye	
Chapter Thirty-Two	**152**
In My Dreams	
Chapter Thirty-Three	**154**
Three Anniversaries	
Book Two	**159**
Chapter One	**161**
Gromit's Tail	
Terminology & References	**167**
Some Zen, Buddhist, Shinto and Huna Terminology	
Influential Books	**173**
Bec's Favourite Books	

Book One

Chapter One
Refuge Savage

I had been a volunteer and palliative care foster carer for a dog refuge in Perth, Western Australia, when I received an email from a colleague letting me know one of the dogs, 'Bomber', was getting a little kennel-crazy after being in the refuge for a year. He had been rejected by all potential adopters, and now, there was only one volunteer he would let near him to attend to his feeding, brushing and walking.

"He is getting very aggressive and as you have Shar-Pei breed experience, could you help and give the troubled animal a break?"

Of course I would! How could I not? It's a dog in distress and how bad could it be? As hard as it was to care for refuge dogs, I missed having those furry creatures around with all the amazing rich life lessons they brought time and time again to my life.

Having prepared at home for his arrival, reinstalled a doggy door, cooked nutritious food, cleaned water bowls, dragged out a soft basket for sleep, unearthed toys from a cupboard, set up an outside bed under cover, I did a check of the fences at my property for escape routes. All seemed in order as I double-checked the sheds were locked, so he wouldn't get in and chew on anything toxic (which he managed to do in another home later in life but that is for another story). Ready to receive this unfortunate creature, I went to the dog refuge that weekend.

I was taken to the dog run, and there he was, a bright-russet coloured, brown and black striped, furry nightmare, with bleeding teeth, a grazed muzzle and ripped ear, literally hanging off the kennel fencing by his torn claws. He seemed in that instant an almost insane bundle of muscle, and he was snarling and foaming at me like he wanted to rip me to shreds.

It is unfortunately normal in a refuge to see many varying degrees of distressed and degraded dogs but I had never witnessed anything like this one and I was quite prepared to turn around as I was thinking, *I won't be able to handle him*. As the full extent of his unhappiness hit me, I felt physically ill and I recall I turned slightly away in denial that any human could have induced this fury and fear in an innocent animal.

There is a reason that refuge records about dogs' origins are sealed to all but a few because there are many who feel so strongly about an animal's mistreatment, that if they could, they would retaliate the same upon the perpetrator so that they would learn how it feels to be treated with such cruelty. The thought did cross my mind as I gazed at poor Bomber.

The wisdom traditions of Zen, Shinto and Huna have taught me that this reaction of retaliation is quite natural but if we choose to respond in kind, then we would ourselves fall into the same state of disgrace and a terrible cycle of violence ensues. Embracing my ill feeling, I looked into the eyes of my new soulmate (although I did not know this then) and asked him silently to let grace arise for just a second between us.

My heart, clearly, is wiser in all decisions than my head, which said, *Run! Run! Run! This will cost you your heart* and bending down to his eye level as he dropped off the wire, I bowed my head and promised out loud that I would do anything that was asked of me to look after him and that love would arrive when we were both ready.

It was an act of compassion and a promise of love yes, but more than that, I felt I was meeting a debt. In the moment Bu's wary brown eyes met my tear-filled blue eyes, our lives were inexplicably intertwined. I saw his savage and tortured nature as a reflection of the untamed and unseen wild in me; my more gentle and calm nature yet unrevealed in him. I felt, too, somehow, I was bound up in the cruelty that had been shown to him, after all, I am part of humanity, and so I was responsible for not only the repair, but also the ultimate grace of the rejuvenation of Bomber's life. These contemplations took mere moments to weave into my soul and more than Bomber's lifetime to realise.

It was agreed that I would have a three-day window to give Bomber a break and considering the potentially dangerous nature of this dog, I was given a muzzle and soft choke collar.

As I led him to the car, he seemed docile and without any sign of wariness. He sprung onto the back seat and curled up to sleep as soon as the car started moving. Obviously, he was truly relieved to be away from all that barking and the cacophony, from dawn until dusk, of excited dogs trying to get attention. It was my dad's belief, when he got to know Bomber, that the noise, on top of the abuse he had already suffered, was traumatising. I think that, as in most things, Dad was right.

Bomber did like the quiet life; he always went away from noise of any kind. Even a movie with loud action scenes was enough to send him onto his bed to curl up and blow hot air into his belly and soothe his furry brow.

When I arrived home with him, muzzled, he simply trotted around the bush block, left his scent where he could and stood at the door to be let in like it had been his home forever.

I knew I couldn't return him to the refuge as agreed. We needed more time. I sat down on my tatty green couch, took off his muzzle, and was startled when he flew up to push me deeper onto the couch, landing on my lap and all but purring, lying there in sheer bliss. We bonded in that very moment.

Every time I tried to move, Bomber would look up, deliberately lean his weight onto his front legs, drape himself further across me and firmly press his paws down. Then, looking into my eyes, he would lower his head back into my lap and breathe short puffs of air in and out, in and out, until the rhythm of his breathing gave way to a gentle snore—adorable of course in a dog, but grounds for strangulation in a human partner!

'BuBu, BuBu, BuBu,' he snored, which put me in mind of the Japanese way of expressing 'Oink, Oink' for a pig's noise. I was glad to be reminded of the years I'd spent living in Japan by this simple sound. And so, this is how Bomber got his nickname BuBu, which of course for me became simply, Bu.

It was the first time I heard his 'happy breathing' and I heard it most days when I patted him or he was in joy or bliss, or simply resting and feeling contented for the rest of his life.

What I had thought would take an immense amount of effort and time to nurture this dog back to any semblance of a happy state of being and to teach him to again trust a human, in fact, took just a little.

My dear friend Adam has a deep philosophical point of view that to enable health we must simply open up a space and create the right conditions to allow flourishing. Adam lives by this and exemplifies it in each and every moment. Indeed within a few short weeks, some anti-depressants and a protocol of natural remedies, including crystal and herbal essences, massage and Bowen therapy, a highly nutritious diet and a bush walk every day, Bu was on his way to repair.

His character changed almost immediately from that of Bomber, the refuge savage, to adorable Bu—my constant companion.

He was not a cuddly animal and, in fact, would happily shake me off or walk away if he had soaked up enough love for the moment. Being petted at times seemed to confront him. A beloved partner of mine, who got to know Bu very well later in his life, pointed out to me on more than one occasion that I had failed to heed the silent message Bu had just put up: *"Too much love, too much love. I gotta get outta here."*

Hearing Bu's silent message to be loved on his own terms, I reflected on a Zen lesson that if we primarily try to shield ourselves from discomfort, we will suffer. Yet, when we don't close off and we face the pain of say a heartbreak or abuse of some kind, we discover a kind of oneness with all beings. It took my encounter with Bu to encourage me to not brush aside what I felt about his circumstances. Instead, I claimed both the love and grief it aroused in me. Most of all, I learnt to not be afraid of the pain it generated.

Just shy of three weeks of caring for Bu, I learnt this never-to-be-forgotten lesson. Buried inside this one big lesson, there were the smaller lessons of not losing courage, of learning to persist, that love is a real palpable force and that it can shift our beings from states of disgrace to grace.

One traumatic day, Bomber bit my then husband, Justin, who was to me, initially interesting and engaging, but it soon fizzled out as I slugged it out daily to support us all. He bit him so hard on his fingers that his nerves to this day are deadened. Being a musician, this was unforgivable. The bite was provoked by Justin taking a cooked bone away from Bu, who was not to know that Justin was doing the right thing because being cooked, it was not safe for Bu to eat. I called the refuge for advice.

The refuge decided to prove that he was not too dangerous for us to adopt and that we should bring Bu back in for further assessment. I could actually hear my heart crack open and break as we returned this poor creature to the refuge. I wanted to run away from there and escape with him but as he was still owned by the refuge, that would have been illegal.

I was in a cold sweat and inwardly crying when I saw the moment he gave up on life as he slipped his head into the waiting collar held by one of the volunteers. He was led to his kennel, taken off the lead, and he promptly went to sleep on the platform above his doghouse with his back turned to me.

I slunk home feeling like a criminal. I had broken my promise to him and sat disgusted and in disgrace in the house made empty by his lack of presence. For three days, I sat doing nothing, silently praying that he would find his way to a home more used to his needs, trying to be calm and to accept that sometimes there were things beyond my control and outside my sphere of influence. With a Shar-Pei and Pit Bull mix to his name, a dog that bites humans and was known to fight all other dogs looming over him, there was a real possibility of euthanasia.

They could kill my dog, I thought. If my heart was ever in pain, it was in that moment of potential loss and so I gathered my wits and called a lawyer who could deal with rightful ownership of animals. In the end, it turned out to be completely unnecessary to go that far; it just took the courage to find someone who could help me with Bu's behavioural issues, I had to sign some indemnity forms and undertake training for Bu for as long as he needed.

Training Bu turned out to be a lifelong commitment. For the rest of his life he fought other dogs and came off badly each time. He was all aggressive noise and no follow-through with teeth. His confrontation would enrage other dogs so they always got a good bite or two in. He would still not surrender bones or allow any other tasty rotting snack to be taken off him without protest.

It took a month to get him back home by which time he was somewhat wasted and unhappy looking. When I went to pick him up the second time from the refuge, I poked my head into his kennel full of trepidation. I actually got down on my knees to see if he would have me back, and as he looked down at me from his sleeping platform, all Zen serenity, I knew he would. Bu wagged his short little tail '*thump, thump, thump*'—always just the three times—he hopped down gently onto the floor and with no fuss, slipped his head through his collar and led me to my car. The kennel manager was there and said it was the only positive reaction to a human he had seen from Bu during his time back there.

I could see all was forgiven; I could scarcely believe the pure grace of the moment. I am still amazed that despite the four weeks' incarceration, a broken promise and broken heart, he had from afar, taught me courage, persistence and the power of love. Bu taught me the pain and pleasure in practicing grace. What an amazing dog, what a Zen master!

My exposure to both Zen and Shinto practice is largely a result of travelling and living in Japan for many years. I had however, read with curiosity about and even tried Zen meditation at a young age without realising the profound philosophy behind it.

Zen, for those curious, is the Japanese variant of the school of Mahayana Buddhism which strongly emphasises concentration-meditation as a way of experiencing a dynamic emptiness. The practice of this kind of meditation is said to open up the mind to a liberated way of living.

Adopting Bu had certainly shown me an aspect of my true nature—I had courage and tenacity and it did feel liberating. I noticed a buoyancy in my meditations that had not been present before, even a glimmer of the depth of calm that is now present in my life.

I believe from the moment we met, I had found my most elegant and closest Bosatsu. A Bosatsu (*bodhisattva* in Sanskrit) is the Japanese word for a being that remains in Seishi (*samsara*) out of compassion. Seishi literally means 'wandering-on' and not, as many people think of it, the Buddhist name for the place where we currently live, that is Earth. In some early Buddhist texts, Seishi was thought of as the process of destroying and creating a place, over and over. A Bosatsu moves endlessly on in search of enlightenment and accruing good Inga (*karma*).

Being rescued and being a rescuer counts as good Inga, for both concerned. Animals are regarded in most schools of Zen Buddhism as sentient beings and as possessing the Buddha seed or the potential to attain Buddhahood, if only with a more limited way of realising it perhaps. But if they are close in any way to evolving humans, then they are considered to possess good Inga and have a better chance at reaching enlightenment. The rescuer attains merit, more good Inga, by the act of their loving kindness.

Bu must have accrued great Inga with, and before, me. He transformed from a snarling brindle beast into a dog possessed of serenity, love and loyalty, and a being who explored all the facets of life with a deep fascination. He softened as he grew into his true nature. I often kissed the velvet fur on the top of his head in wonderment that I was invited to be so close. Sometimes he would put out a paw and place it on me to stop me as I passed by, then he would graciously accepted the gestures of affection that came...a stroking of a furry belly, a tickling under a chin or a whispered sweet nothing in a tiny ear.

We knew love and compassion in each and every minute we spent together. We knew the profound transformation of melancholic hearts to joy filled ones. We knew the utter bliss of living a life fully embraced in and of the other.

Bu's Lesson—Every being can rise from disgrace to grace. Appear by their side, press up close, extend your paw to help and they will reveal themselves to you.

Chapter Two
Palm Shimmy

Bu loved to be outside and spent many happy hours patrolling remnant bushland and the fence line of my one-acre property in the Perth Hills, scaring off possums, kangaroos, rabbits, stumpy-tailed lizards, large frogs and rats.

The rats were quite invasive at times and lived full lives, hiding out in the wood shed, the pump room, the roof of the house, under the house and in the palm trees that someone had inexplicably planted alongside an enormous oak tree that shaded the entire backyard. Perhaps palm trees were fashionable in the era in which the house was built, about a hundred years ago. Now they were the perfect hiding place for the rats, a place from where they could tease Bu.

For months after Bu and I arrived, I would see him sticking his nose into corners, snorting out his prey, derrière and tail waving with anticipation in the breeze and looking surprised he hadn't caught any. He must have gotten frustrated with the rats hiding in the palm treetops—what a challenge!

One day I looked up from what I was baking at the time to see Bu, out of the kitchen window, scaling the trunk of one of the palm trees. Up he climbed using what I can only describe as a shimmy…all four of his legs were gripped around the circumference of the trunk, first the front two paws would rise and grip, then the back two, and then he inched his way to the top fronds where his prize was nesting.

He was making pretty good progress until about halfway up the palm he encountered the sharper, fresher ends of the fronds cut close to the trunk. Becoming rapidly unstuck and falling away, he made an ungainly dismount that ended in an 'ooff' with his face planted onto the ground, not for the last time in his life. He proceeded from the ground to look at the palm with dismay, and then, it was about-face and a disdainful trot off, tail held stiffly erect in the air, as if he had better places to be, which he did, at the back door, waiting to get a bit of whatever I had baked.

He never, to my knowledge, tried the palm-shimmy again but he would patrol the base of those palms with a daily vigilance that was heralded by a peculiar set of high-pitched short woofs as if to say, "*One day you wily rats, one day.*"

A winter creek ran across the property and when it flowed, Bu entertained himself by running back and forth along the length of it, trying to root out those noisy frogs.

There were three sorts I identified and they all drove him wild. Together they created a maddening mix of alternating, long, drawn-out, mournful moans such as the '*hoooooo*' of the moaning frog usually late into the night, the metallic-rattling of a 'pea in a can' made by the clicking froglet throughout the day, and on and off both night and day, the explosive '*bonk*' of the banjo frog.

It made the setting of my old, weatherboard cottage in the hills seem idyllic and romantic, only ever interrupted by the arrival of the postie on his bike. There was something about the sound of that two-stroke engine that set Bu's hackles upright and he would rush from an alert state of frog patrol or rat hunt to running up and down the fence line, barking furiously at the bike with every sign of enjoyment at startling the enemy riding on the other side of his property. It was fortunate that Bu was never a digger or a jumper, as it was a pretty low fence and not that secure despite my initial checks. It was just enough really to outline where his property ended and the main street started.

It never seemed to me that Bu pursued happiness. He was simply in the moment, as if he were a practising monk, devoted to being aware and to cultivating mindfulness. My practices of Zen, Shinto and Huna all teach connectedness to one's own life and to experience and focus one hundred percent on what is present, not what is missing.

Ending futile quests to restore us to a perfect whole or blissful happy state is seen in these traditions as the path to peace and contentment. If we let go of what is perceived as deficient in our lives, and the longing for things to be other than they are, we will surrender and can be released.

All his life, Bu was one hundred percent focused on one thing and then he would simply slip into being one hundred percent focused on the next, which for him was normally an active hour of rat, frog or motor bike pursuit and then, happy and exhausted, a solid two-hour nap.

I found it endearing, admirable and exemplary. The days I emulate this practice are the days I find most peaceful, as no matter what is going on, be it good or bad, I am aware I'm enjoying just being in the moment.

Bu's Lesson—Whether a rascal possum chaser, impossible palm tree climber or perfecter of long doggy naps, give one hundred percent to what is and who you are – not what isn't or who you think you should be.

Chapter Three
The Poo Connoisseur

Oh, how Bu loved poo! He loved all kinds of poo. It was a source of constant delight for him to supplement his menu with the more, shall I say, organic findings in the bush. To him, discovering these nuggets was like finding gold. Make no mistake, *he was a very well-fed dog indeed.* While I perhaps had some flat soda water, a slimy block of tofu with maybe some wilting vegetables in the fridge, his section was packed with nutritious home-made food and *always* a roasted chicken to hide his medication or to give him a treat.

But poo was a favourite fare. The times when he struck gold he was like an excited prospector. He would pause in mid run, turn on the spot and press his nose suddenly and hard against the discovery. He would snort like a pig and drop down on his haunches to feast.

He would crunch his way through a handful of rabbit droppings and discover that he could, in fact, scoop up quite a few of these and munch on the run. *So many trees to sniff, no time to stop!* Kangaroo pellets were this connoisseur's delight; he would give off a little keening noise and tuck in with every appearance of a starving waif at a banquet.

In a paddock full of cow manure, he was spoilt for choice; though he preferred it perfectly dried like a cracker. In fact, he used to wander into my friend Bruce's farmhouse in Gidgegannup and steal the dried cow dung used in the Agnihotra ceremonies my friend held, which were set just outside his house. Bruce himself is an incredibly kind, earthy, wild and loving human being and his face always lit up when he saw Bu and scolded him only mildly for the raids.

His Agnihotra ritual uses dried cow dung 'crackers' burned in a copper pyramid, ignited with ghee, and accompanied by a small offering of rice. The rituals are held at both sunrise and sunset for purifying the atmosphere through specially prepared, healing fire. During the ritual, a short mantra is chanted which is then followed by a brief, silent meditation.

Bu and I sat through many of these peaceful meditations and despite his poo-cracker theft, there was always plenty of the required materials, so his misdemeanour was overlooked with amusement and he was always welcome. Mid ritual, Bu would gently nod off, begin to snore and remind the meditators to keep the inner eye on self and not the distractions of a hound with a full belly, who was dreaming of hunting rabbits and playing cow stampede.

Cow stampede was a game Bu invented with the cows on the farm to keep himself entertained when we visited my beloved Tarryn, who lived just a short walk from Bruce's farmhouse. Let me diverge here slightly and share who Tarryn was to me. A long, curly haired, lean, six-foot-two man with striking Viking features—piercing eyes of a startling blue, I was utterly enchanted, like I was under a sweet spell for the time we were together. There was a very strong attraction between us from the start. The first time I met him, I was taken by how exotic and alluring he was, and then years later when we met again, I was caught fast when he turned to me and said, "*I find you intoxicating*,"—what woman could resist that! We shared many travels, adventures, spiritual explorations and a love of Bu. We still enjoy a friendship but the romance, being hot and fiery, burned us both and suffocated in its own heat and eventually we let the suffering for love go.

Cow stampede went like this: around sixty or so cows would wander close to the paddock fence where Bu, by Tarryn's caravan, appeared to be sleeping in the sun. But Bu would suddenly be up, furiously barking, and rush at them, and they would stampede back, seemingly in mock alarm.

The cows would stand and stare at Bu and then, as one, they would gather speed and come thundering back towards him, diverting away at the last minute, making him run away. The game would be repeated a few times until canine and bovine energies tired. Endless hours of fun were had playing this game and when Bu arrived at the farm, he would generally run to the fence to see if his playmates were there, and if not, of course, there was a paddock to explore and fresh cow pats to savour!

Bu also fancied horse manure and as there was only one horse on the farm, Rosy, her droppings were a little hard to come by. When Bu did find some, he would do a funny little dance up and down on the spot before tucking in with every sign of consummate pleasure.

Bu had only one close up and personal encounter with Rosy. Tarryn and I would feed her apples and carrots when she came by to see if anything was on offer, and on this day, Bu had been very well behaved and had even left alone Bruce's scores of free-range bantam hens, so we decided it would be fine for him to say hello to Rosy too.

Rosy lowered her beautiful long face down to meet him and Bu stretched up, paws resting on the fence rail and he sniffed her breath as she whinnied in greeting. He wagged his stumpy tail and looked thrilled to make a new friend. He jumped off the fence, executed a tight circle and then jumped back up again. Rosy bent down to nuzzle him, clearly comforted and was most indignant when Bu took hold of a sizeable chunk of the right side of her lip between his teeth and stretched it out until it snapped back in place when she pulled away.

He was not in my good books for the rest of the afternoon as I contemplated the issues of trust that I had with this creature and what I felt was my anger. I was angry at two things, primarily that he had attacked Rosy and secondly, that although I excused his action as arising from him being brutalised in his puppy years, with all the love and care he now had, he still felt the need to attack. It wasn't only that he'd attacked a quiet horse like Rosy, but that he attacked almost every dog he came across. I was also angry that Bu had been treated so badly by some humans and it was playing out in his life so strongly.

As someone learning the lessons of Zen, I try to respond with equanimity, with clarity of thought and not out of my accumulated habits and messy emotions. *What should I do about this,* I asked myself and then, I began to think that actually…this is the wrong question…what I should do is not important. What I *am* already doing is important. Observing my thoughts, I watched the feeling of anger simmer. I didn't try to stop the anger and I didn't tell myself I was wrong for feeling this. I took some steps back and examined the reactions born out of the habit of trying to escape from the reality that dogs do get treated inhumanely, even as they are loyal and loving.

This situation sometimes causes wrath to arise in me and it has become the catalyst for me donating my time and energy to organisations that rescue dogs and encircle them with all the powers of justice they have to protect and love them. I am inspired to be part of these compassionate organisations and participate with all the resources I have.

I now ask, even when I feel angry, what has happened to these people who feel the need to perpetrate such crimes against a living creature? Perhaps their situation was as horrific as the damage they met out to these suffering creatures, and as difficult as it is to contemplate—perhaps their situation was possibly worse?

A learning I received from Shinto is that every person on the planet is intimately connected. Shinto teaches that even acts of violence or harm against another creature are expressions of the universe, and, as part of being human, we have in some way, been a part of this. As all forms of violence cause suffering, for both the perpetrator and the receiver, the lesson is that we harm ourselves when we harm others and we will find peace elusive should we persist in violence.

Non-violence is one of the many essences of what the Buddha taught. Its practice is liberating in each and every moment and suffuses one's mind with compassion, identification and empathy with other beings. My own interpretation is that if we truly believe that qualities of heart and mind constitute our enlightenment, and that the highest welfare for all beings is a life of harmony and peace, then permitting someone else to perpetrate harm without consequences is not non-violence—it's simply pacifism. Spiritual practice means sometimes living in a state of paradox and so I watched Bu calm on the dried grass and loved him as he was in all his states—as the Rosy attacker, the rat catcher, the frog hunter and the connoisseur of poo.

So back to the poo. One day far from the farm and the bush, in the middle of suburbia, an animal circus had come to town and camped on a large oval where I occasionally walked Bu. Oh boy, did that dog hit the jackpot the day he discovered elephant poo. He must have felt like the luckiest dog alive, beside himself with doggy joy. I swear he did a double take walking past the enormous pile of dung waiting for him to savour under the old olive trees. I think if he could have, he would have taken some home with him…in err…a doggy bag.

I was never disgusted by Bu's peculiar love of poo because it was vegetarian derived and perfectly safe for him to eat, besides, he seemed to take such delight in its finding and eating. Although he mostly snacked on it in my sight, there was one day when he had taken off into the bushland in the South West of Western Australia and had probably eaten fox scat; whatever it was, it made him extremely ill. When it turned his bowels liquid and he could barely hold himself up, I took him to an emergency vet clinic. After a couple of injections to target various possibilities and make him a lot more comfortable, he was allowed to go home with me, armed with cans of bland food designed to be gentle on his stomach.

Bu was not impressed when first offered this fare and was a very sad and sorry sight for a couple of days until he bounced right back, his nose twitching and tail held erect in air, as he went in hot pursuit, once more, of that most sacred of his food treasures, poo.

Bu's Lesson—True connoisseurs know that sometimes you take a bite out of life and sometimes it takes a bite out of you. Sample everything laid before you, maybe not poo – that is just for dogs?

Chapter Four
Nannup Tiger

Just outside of Nannup, a town in the South West of Western Australia, was Bu's favourite walk in the whole wide world. Driving down south was fun, for what dog doesn't love hanging their head out the window and woofing at the sheep, the cows and the horses as they speed by. He would sleep in the back seat to about half way and then I would say, "*Moo Moo*" and up he would get, thrust his head out the window and race right to left, left to right, not knowing which was the more exciting side to stick his head out. So many delicious country smells to take in!

He was very loved by the local population of Nannup and he received endless pats as we walked the streets to get to one of many lovely local lunch places. He was even allowed in the Community Resource Centre (bypassing the 'No Dogs Allowed' signs), courtesy of other dog lovers.

My great friend Gary, a huge bear of a man with a heart as big and brains to match, lived a little way outside the main town. He was someone Bu had a huge soft spot for. I used to joke with my friend that the reason Bu would lick him with such excitement around his bushy bearded face must be the smeared meat paste behind his ears. Bu was genuinely happy when Gary came to visit, or when we would head down to Nannup to stay, as it allowed him a free run of the surrounding bushland. Streaking through the undergrowth and between the young jarrah trees, Bu looked like a thylacine, a Tasmanian tiger, once believed to have also existed in Nannup but now, alas, extinct.

The Tasmanian tiger was believed to have become extinct sometime in the 20th century, when the last captive one died at Hobart Zoo in 1936. There have been a large number of unconfirmed sightings in the South West over the years, with tales of the famous Nannup tiger still circulating. The only known existence of thylacines in this region comes from bones, estimated to be thousands of years old, in the Mammoth and Ngilgi caves in Margaret River and Yallingup, (also towns in Western Australia's South West). I think of Bu as the last Nannup tiger.

In Zen Buddhist teachings, life is represented in many ways and often, as a journey in which we cross rivers, toil through deserts and scale high mountains. Kind teachers advise us to use only the assistance we need for each stage –a boat is left at the shores of the desert upon crossing the river; long robes to prevent the sting of the sun are left on the desert sands; and when setting foot at the base of a mountain, commence with very little. We do not carry beyond what is required. These things would not only be superfluous, they become a burden. In the Buddhist journey analogy, the series of animal extinctions are not the failure of divine or human plans and interventions, but simply, what is required to move through succeeding situations and environments. In the life of the planet, one group of species was needed at one stage, but these were then unneeded and became extinct.

I am personally appalled at the rate of extinction, human expansion being a proximate cause of it, and I struggle with a philosophical stance that accepts this. But as with anger, I choose to be an activist to face the threat and mindfully so, as an act of compassion. There is a particular path to tread as a modern Buddhist which looks like living a paradox; to lead a deeply spiritual life and yet be an engaged social and environmental activist. Perhaps, each of these lends strength to the other.

Bu's Nannup bush walks always renewed my enthusiasm to embody this curious paradox, becoming a 'sacred activist', which is a term from spiritual author, Andrew Harvey, a magnificent speaker and someone I find inspiring to read. My thoughts about sacred activism always started where Bu would casually wind his way up the fire breaks in the bush, executing figures of eight faster and faster down the gravel track until he would shoot out to one side and off he would run. His brindle stripes provided perfect camouflage in the terrain.

He was never a stealthy animal and really did bulldoze his way through life and the Nannup bush, where we would lose sight of him quickly and often. At times, we would be calling out until we tired of waiting, resigned ourselves and sat in the car, poised for him to come back. There were other times when one of us would go and get breakfast, enjoy the food and wait for him to return happy and tired.

There was one occasion when I almost made myself sick with worry. We had called and called and as poison baiting for foxes was done in the area, I was trying to figure out if the State Emergency Services would send out search parties for dogs. There Bu was though, sitting at the car, patiently waiting for us, innocent of all bad behaviour.

Worry and fear are strange emotions that can drive our entire lives if we let them and they can certainly make us sick; we see the future through a distorted lens. It is legitimate to experience fear, yet the distortion focusses us on the negative outcomes and will cloud the positive alternatives. It can take the life out of life; when Bu did not return quickly to a call, I knew what 'having your heart in your mouth' felt like. To carry this around while walking in the bush is exhausting.

Being mindful of the fear and recognising its appearance, we cannot be imprisoned in a tense body and be possessed by an aching heart, we can set ourselves free. The presence of fear may be the result of an accurate perception as well as a completely skewed one, but either way, we can control the level of fear we experience.

Missing Bu for an hour or two, I learnt again, all things change constantly, even what is most precious. I know that I and those I love will die. Contemplating a life without Bu was both the fear trigger and the greatest of teachers to cultivate mindfulness and release; creating purpose and avoiding suffering. I knew that Bu himself was never lost and never in fear as to finding my location!

After his bush expeditions, Bu would curl up on Gary's couch and snuggle into his blanket to catch a long snooze. The right end of it in summer as it was hot and he could catch a breeze through an open door and the left end of it to enjoy the warmth of the potbelly stove in winter. He was a dog that sought comfort and getting to the right temperature was important.

In my Perth Hills home, he would stagger in from the sizzling heat outside, where he had been laying on his outside bed soaking up the rays, and then flop down on the lounge room floor to cool down, only to get up again cold and head outdoors. Bu did not suffer mindlessly, he navigated his life from moment to moment and used exactly what he needed and no more to do this. Moreover, he did it without fear.

Nannup tiger style, my Bu leapt into thin air one morning, this moment indelibly printed on my mind as an act of defiance against what a reasonable sentient creature would do, against gravity itself and the frailty of an ageing body. He sailed off the deck of a friend's house, about eight feet above the sloping land, after a kangaroo. In the suspended moment of shock, his doggy mates Mini (who he quite fancied and would sit next to with his little tail going round and round in short circles—it was unrequited love as she really wasn't that keen) and Me (another dog) gazed after him in disbelief with the rest of us.

He landed heavily on his chin and chest with a loud 'ooff' (again) and ground to a graceless halt but shook leaves and dirt off and set out at a great rate to enjoy the chase of a fast disappearing, grey brown tail. Bu owned each moment and taught me that this kind of leap, in the less physical sense, is something that scares us and draws us at the same time. No matter what we do or where we go, we must cross thresholds despite fear and trembling, face certain challenges no matter win or lose, make seemingly unbearable choices, and find ourselves lost and without a map to circumnavigate the globe of human experience and yet, sail all the same.

I take a leaf out of Bu's book and try to own the moments of joy, of sadness, of ecstasy and anxiety, and leap off the edge on occasion. Perhaps the discovery beyond will be worth taking the courage, and even if not worth it as an outcome, pleased, I still took the chance.

Bu's Lesson—Go on adventures, taking only what you need to make that leap, big or small – you may discover you can fly!

Chapter Five
Bomber's Origins

I once lived in a beautiful limestone house, reminiscent of a cathedral, in Fremantle, the port city south of Perth. When I decided to sell up and move back into the Perth Hills bushland, I engaged a lovely real estate agent, Lynne, to help me sell the property. When Lynne came to the house for the first time, Bu of course followed me to the gate to make sure there was no threat and he could execute his guard duties if required. He got petted and wound his way around the estate agent's legs like a cat. He was normally very stand offish with strangers so I was quite surprised. "*Bomber, put her down*," I said. Lynne looked up and said, "*It couldn't be...it looks like him, my friend had a dog called Bomber, but had to give him up for adoption.*" Being a very difficult dog, the friend had tried three times to adopt him out but he was always returned within in a few days because of his severe behavioural issues, until her partner said he would take care of it for once and for.

Lynne bent down to Bu and said, "*Hi Bomber!*" He madly wagged his tail, causing his whole back end to sway in delight, a thing he did when he was a puppy apparently. She relayed to me that her friend's daughter called him "Jo Jo Bomber" and he was born of a dog named Sabra who was a Staffordshire cross Bully cross Kelpie cross Dingo cross. Lynne put me in contact with Anita who wrote me reams of information about the short part of his life she had had with him and his disturbed but irrepressible character.

It turns out someone in a long line of homes had re-homed him poorly and he had fortunately ended up in the refuge, rescued from severe trauma and abuse. Anita told me that Bu was an only pup, born on 17 July 1999, and had been sadly rejected by his mother Sabra. Anita was happy to share some "tails" and escapades about Bu when I told her about the love I had for him:

"Sabra was reluctant to let Bomber feed and kept flipping him away from her. If dogs get post-natal depression, then she had it! She wouldn't eat anything but sausages straight from my hand and used to tip her water out of the bowl as if she couldn't bear the sight of it. She was extremely confused and unhappy. I had to put Bomber onto her teat to feed and the only way I could get Sabra to stay with him at night was to have them both on the end of my bed. I actually forced her to look after him. As he got a bit bigger, he would push two of Sabra's teats into his mouth at once and grew enormously fat. No wonder he thought he was the King. He was like a balloon with four paws.

"Despite being rejected, Bomber was a very happy little puppy. He was beautiful! I used to carry him around in a baby sling and he was very much loved. He was a very 'wiggly' dog and when he wagged his tail, he wagged his entire body so that his nose touched his rump either side. He was very different to Sabra and didn't respond in the same way to any attempts at training or socialising. I'd never had any difficulty training Sabra because she was very sensitive and loving, but Bomber just wanted to take over the planet! He would get up onto the bed and make a point of trying to push me out of his way—just to show he was boss! He was so clever, he used to trick me at every turn! And what an escape artist! He escaped from a cage at the vet when I took him to be sterilised!!!"

Anita also said he often escaped, stole shoes, barked at people to scare them and loved high places; even getting stuck on house roofs. It all resonated with me and the experience I was having being Bu's carer. I was astonished to have met Lynne, then Anita and have Bomber's origins revealed to me. Even more amazingly, I met the vet she talked about a year later through another good friend, Jay, a joyful, thoughtful and handsome environmental activist I have known since high school. On meeting Bu again, the vet was reluctant to pat him, saying she had sterilised him and had recommended to the owners to be very careful with him as he was highly aggressive and uncontrollable. I knew he could be like that too; it was simply part of his makeup when triggered by certain events and situations but never with me.

I could call it coincidence, luck or even fate but I know it is so much more than some random occurrence which is devoid of any deeper meaning. These encounters were delightful and I enjoyed them. As Zen practitioners discover, coincidences will seem to happen more and more often when on a spiritual path. They arise from real connections which are only known by intuitive awareness. We would have to be fully enlightened to really understand them, but we can navigate our way through life by recognising the auspicious signs of these connections as they manifest in our lives.

Psychologist Carl Jung described coincidences as events of synchronicity as 'the underlying cosmic intelligence that synchronises people, places and events into a meaningful order'. He describes them as the 'magical effect' archetype; a trait universal to all human beings, that is innate and intertwined into the collective unconscious. I looked at Bu every night since I learned about his life in his puppy years and was awed by a sense of rightness. From the time I saw him hanging off the wire of his kennel, to all the heartache of his illnesses, to all the fantastic adventures and all the love shared, I was meant to have my life entwined with his, to learn and to be open.

Lynne did sell the house for me, in a rather bad market, she was diligent and of course, caring of Bu when he was left at home with her to open or close the house for inspection. One day, she arrived at the house and I got a frantic phone call that she could not find him anywhere! She had looked in all the rooms, the garden and called out for him. I immediately made my way home, there was a thunderstorm on the way and he did not like those one bit.

He would tremble and whine and cower and be quite unlike his usual warrior-like self. Halfway through my travel home, I got a call that he was okay, he had hidden in the bathroom and the door had locked behind him. I got home to a frantic dog who, in his haste to get out of the locked bathroom door, had gnawed his way through a good portion of wood and lost a front canine tooth in the process. This of course led to a trip to the vet to get it looked at, removed, and the gaping hole left behind by the proud canine tooth was stitched closed to let it heal.

I was not surprised, he would do anything to escape thunder and lightning. He once had eaten through a locked metal cat flap; it looked like someone had taken a can opener and neatly cut it all around and, of course, being dog size, he had widened the gap by chewing the wood around it to escape outside. Another time, he munched on concrete around a gatepost to get out and roam and seek comfort but returned home calm and in control. In other adventures, he scaled a six-foot fence, hid under beds and chewed on the supporting iron posts. He was truly afraid of the sky falling on his head!

Once, he even hid in a fireplace full of ash to get away from the noise and find some comfort from the cacophony. I would sit with Bu, calming him by singing, and stroking him, whispering silly things into his tiny ears and generally make a fuss until the thunder rolled away from us, out of hearing range. I would even curl up in his basket around him when he would let me; an intimate friend in the hour of terror. I loved the chance to demonstrate this intimate compassion. Intimacy with another, such as Bu, nourishes us and surfaces the best in us. When the closeness is unbearable, this is intimacy challenging us, (too much love I gotta get outta here) and it may surface the darkest parts of ourselves, those shadow traits that we should lend no energy to. Rather than flee, perhaps intimacy can evoke instead, the presence of our best self-potential.

The extraordinary connection with Bu I feel brought out my best self and continues to do so, as I now look after three more rescue dogs; Diesel, Stella and Jessie.

I do not have the same depth of intimacy with these three beloved creatures yet, but the heart that learnt to be open and be in intimacy with the rhythm of another's life, observant of synchronous events, is still beating very strongly. Bu truly was a Bosatsu in both his life and death; teaching me that cultivating intimacy motivates and inspires us to develop deeper self-knowledge, without which, I feel love would surely perish.

Bu's Lesson—It's no coincidence that you'll find a friend to love who'll love you when you are afraid, hold your paw and stroke your fur until you feel better – they'll feel better too as friends' love is needed.

Chapter Six
Entertain Me Boring Stooge

Bu was a self-contained animal and preferred to be left to his own devices. He would only be truly present in the same room for hours if it was winter and his bed was placed under the air conditioning vent so it was warm. Of course, for a bored human, he was a source of captive entertainment and without actually annoying the poor creature, he would be the boring stooge for my pleasure.

I loved playing with the fur that peeked out from between the pads of his feet and tending these would give him a little massage with the brush as he often got itchy feet from his bush walks. It felt nice to him and I know this as he would fall asleep snoring 'BuBu, BuBu, BuBu' within moments. This was curious to me as it was the only time, he ever let me touch his feet. In other circumstances, he would grumble gently, an improvement on the first time I tried to touch his paws and he nearly ate my face off, quickly snatching his paws away from being touched.

On 'make your own sushi nights', Tarryn and I would put aside a couple of Nori sheets just to watch Bu salivate over them, only to try to pick them up and have them stick to the outside of his lips. He would spend the next few moments getting crazy, trying to lick them off, even as they dissolved into a soggy mess, which he still would gobble up. Hungry for a snack, I would sometimes make a peanut butter sandwich and lure Bu to my side, with a corner of bread laden heavily with peanut butter. Drooling, he would take his treat and begin to eat and end up chomping away at the sticky mess in his mouth, moving side to side and gyrating his furry behind up and down in time with his mouth as his tongue did gymnastics in circles around the glob of peanut butter. We would laugh like mad things at his antics and even harder when he had made his way through this and came back expectantly for more.

I love to entertain and am a rather passionate cook, so it was not unusual for friends to be over for various meals and sometimes exploring of new cheeses, French butters and creams, all piled up onto huge platters with an abundance of fruits and crackers. I learnt once never to leave these unsupervised, as an opportunist dog can consume many layers of multiple diary delights in the time it takes to walk to the door, greet and hug friends and return to the table—to find a whole stick of French butter, a whole wheel of brie cheese and a tub of clotted cream gone from the platters. The remaining evidence—a set of smiling, greasy doggy lips.

It would have been easy to be mad at the theft, but I laughed so hard at the expression of delight from consuming forbidden pleasures, I almost hyperventilated. I recall my friends were in stitches, watching the butter grease drip from his face and his tongue making slurping noises as he cleaned off what he could. When we recovered, we decided to tuck into the remainder of the goodies, ignoring the slightly depleted selection and the ragged appearance of some of it. I think we got fifty-five dollars' worth of entertainment from the event in any case and of course, a story to illustrate why a dog, a low table laden with treats and a three-minute window of chance, is a poor combination.

We took Bu on a forced march (he seemed somewhat disinclined with a belly full of stolen cheese and butter) later that evening to make sure he would walk off all those calories—he was waddling slightly as he proceeded in a stately manner around the nearby lake, making no attempts to see off the ducks roaming the grass right in front of him. He would rarely pursue them into the water as he hated getting his feet wet. Imagine his surprise when he trotted jauntily around the rocks in a lovely local bush reserve which had a fast-flowing waterfall that would rush up the rocks, and as he was standing and staring off into the distance in deep contemplation, he got swept away.

No matter how hard he scrabbled, he could not get any purchase on the algae covered surfaces. By some act of grace, instead of being carried along bumping and being mashed against the rocks, Tarryn reached down as he went by and literally picked him up by the collar and tail and swung him to safety. I think he got even more cautious about water after that.

He loved his walks and Tarryn and I could not help but tease him each time and be entertained by his responses. We'd crouch down, lead in hand and say, "*Do you wanna go for a... helicopter ride?*" and watch him sit up. "*Do you wanna go for a...quiet stroll?*" and hear his breathing get heavy. "*Do you wanna go for a...bungee jump?*" and watch him stand up and look at us. "*Do you wanna go for a...W.A.L.K?*" at which point, he would be beside himself with joy and park himself at the front door in great expectation. He would always sit very calmly in the car on the way to a walk like he was contemplating the divine in the moment of each experience.

Calculating the average number of walks we took in the eleven years of being together, at an average of two walks per day for eight years and then one walk per day for the last three, we took six thousand nine hundred and thirty-five walks, give or take, being more when we visited the Nannup bush and less if it was pouring rain and there were too many puddles to avoid. I found the walks themselves mostly fun and geared to fitness but I also enjoyed them for the observation of the experience of being in nature, the incredible changes that the seasons would bring and of course for my beloved pooch.

I am never bored walking in the bush and there was never a moment of boredom with Bu in my life, his diverse character, the entertaining events and the pure life lessons filling the hours. In the months after his passing, I had many lonely hours, surrendering to the lash of boredom. Each time I tried to stop from doing and tried being, I would spend just a few minutes in meditation and then, I would be up baking, taking a walk, napping, reading, putting music on, going out to movies, checking in with friends; too much of any of these would of course lead once again to boredom. Anything to not just fill the void left by my soulmate's departure but to also alleviate the boredom of the cacophony of sorrow I was generating. I could have let myself become consumed by the inertia of grief and the weight of the nothingness.

The bruised ego suffering with the loss of love; it played so many tricks on me, this ego of mine—did I hear my phone ringing? Did I need to ring my mum? Maybe I should get another dog to help heal? I wonder what I should cook for dinner? I needed to make a bold move to encounter my unhappy mind.

I learned from Huna to embrace the boredom and cultivate instead past remembrances of the feeling of the stillness, especially of the evenings I had spent playing with the velvety ears of Bu. Eventually, I got bored with contemplation of boredom and emerged into the landscape of moments of lightness. I began to have clarity and treasured the relearning of the intimacy with myself in the space of meditation. Recalling the serenity of an evening observing a rising moon and listening to the eerie quiet a heavy mist in the Perth Hills will bring to the bush was my entertainment.

I tend to seek distraction more mindfully now, consciously engaging in play with my rather large rescue hounds. They are a combined weight of one hundred and twenty kilos and take a lot of care, which is joyfully given.

Despite their somewhat daunting size and constant demands, there is time for reflection and meditation; time for looking inwards. Looking inwards takes time to do properly, for me at least, and a commitment to practicing meditation. Most days, it's pretty easy, even with Diesel, or Stella or Jessie sitting with me in the meditation space; I guess they like the vibes too and within moments, I hear their snores…the boring stooge is not entertaining today.

I'll be honest and confess, it's still very funny to watch a pooch eating peanut butter and enjoy the look of anticipation on their sweet faces as I reach for the jar and scoop out a tiny portion for each of them.

I do try not to make them stooges to entertain me if boredom strikes; a very rare occasion indeed now. Life is so full and there is something infinitely intriguing in learning about living in each moment to enforce the lesson to look to that which is seeking and searching. Myself.

Bu's Lesson—Take a walk with your dog if you are bored, then they won't eat your shoes, dig holes in your garden or steal peanut butter out of the pantry – and you won't be bored either.

Chapter Seven
Full Moon Drumming

My dear friend Bruce had built an enormous and beautiful tipi on his parents' farmland out at Gidgegannup. The wind would sigh through the trees and all day we would hear the cry of the red-tailed black cockatoos, the eerie squawks of the straw neck ibis as they settled in their hundreds in the skeletons of the marri trees, and the gentle cluck of the chickens running around scratching the ground for bugs. I found the bantams hilarious in their feathered pantaloons and so tame, they would take seed right out of the hand in rapid little pecking motions.

For a dog with a very strong hunting instinct, Bu to his everlasting credit, left these girls in their natty kicker elastic trousers, completely alone. He seemed to know that he would be in utter disgrace if he even so much as touched these birds. Bruce had a dog Jemmy, whose duty it was to protect the animals on the farm and clearly, she loved Bu. She was one of three dogs he did not attack on sight and had an enduring friendship with.

As soon as I would drive up to the farm, Jemmy would bark her greetings and rushing out, run beside the car until I slowed and opened the door so she could hop in and cover Bu's face with a few licks and show her excitement with much wagging of tail.

The tipi to this day is still the heart and the soul of an extensive loving tribe-like community. Folks travel far and wide to attend various events, but none are as loved as the full moon drumming evenings. Hippies, self-proclaimed shamans, sceptics searching for more, and people like me, sitting somewhere in between the mundane and mystic, all gather to honour the full moon and sit in song. Drums, flutes, didgeridoo, clapping sticks and Indian harmoniums meld together to invoke the Goddesses' blessings. The music pours out to invite them in. Sometimes, sky-clad, people would revel in the light of the full moon and offer prayers and blessings to emptiness and everything. I don't know what Bu and Jemmy made of it all, but they both seemed very content to meet and greet and show everyone to their places.

Bu was always welcome in the tipi and his place was curled up in a tight little ball in front of Tarryn and me. I would be nestled back against Tarryn's lean muscular frame, our long hair intertwined as if seeking union. Sometimes we'd be so close to the fire in winter that the smell of slightly singed fur would greet our nostrils and mingle with the sharp scent of burning sage leaves and the sweet smoke of Paulo Santo. In summer, Bu would lay stretched out behind us, close to the lifted sides of the tipi to catch the rare breezes and stay cool. A dog has to have his comfort!

Bu and I often stepped out of the tipi for some fresh air and looking up at the sky to take in the beauty of the moon and the twinkling crystals of stars. I fantasied that he might become one of them, reflecting that in the Huna tradition stars are used for predicative guidance. Maybe my Bu would be up there someday guiding and helping. A constellation, as Huna teaches, is to navigate life by as it supposed to be lived – without harm and in service. In these moments of deliberate pause, taking slow breaths, keeping a soft and steady gaze on the quiet of the rolling hills, the noise of the full moon drumming drifting out from the tipi would fade and peace would suffuse my whole being as I stood there in the silence of my own mind.

Bu was more than content to be at my side, so he would retreat too and seek solace in the quiet compound where he would sit on the steps of Tarryn's old caravan, a gypsy dog looking up at me and out at the world with Zen-like serenity.

Now, when I look up and catch sight of a full moon from my window, hanging there in the majesty of the constellation of stars, I count Bu amongst them, the brightest of them all. A shining Bosatsu that unexpectedly continues to guide me through my fortunate and love-filled life.

Bu's Lesson—Step outside, look up at the moon and let its light reflect the beautiful galaxy within you – it is a magical and mysterious place full of wonder.

Chapter Eight
Inorganic Snacks

Bu loved to snack, he was a true connoisseur of poo, but he did not limit his appetite to the organic world, his tastes ranged far and wide to incorporate those items which simply should not be classed as food, for anyone. He cheerfully consumed what would amount over his life to about forty screwed up, used tissues. He retrieved these from the bathroom bin, having taught himself to open it by stomping his foot on the pedal and quickly shoving his head into the bucket to hold up the lid. He would fossick around and then toss his soggy finds out onto the bath mat and chew them slowly to bits.

He was a frequent visitor to the kitchen bin and contents of the pantry. The dog could open any door, be it a lever style or round doorknob, so no place was truly inaccessible. He was intelligent and very curious about things, especially things that might be edible. His definition was wide and varied, don't forget, of what was consumable. Foam trays were of real interest to him and although I am pretty sure he only nibbled these for fun more than fibre, I did get worried that explaining to the vet why my dog got hold of these things in the first place would become embarrassing and I would be labelled a negligent mother. I learned to put things behind doors and up so high that there was no way he could get them.

He once devoured an entire six-pack of plastic wrapped tissues which I had put carefully on top of my handbag to remind me to take to work. It only took him the time it took me to take a shower, dress and run out the door to work his way through these. I was impressed, at least they weren't used, at least he'd stripped the plastic off and eaten them all instead of leaving soggy bits of tissue all over the bathroom mat. I figured I would need to take him to the vet for some expensive procedure, but he showed no signs of distress when I returned just a couple of hours later to check on him.

Concrete was no particular barrier to him either. He happily chewed through the rotting concrete around a post holding a gate up to get out and sit and wait patiently for me to get home from the hardware store, which funnily enough, I had gone to buy more concrete for another project around the home.

As well as chocolate, Bu loved the foil it came in. I know dogs shouldn't eat chocolate, but he was a sneaky thief of whatever he could get his paws on and he caught me unawares more than once. All I would find was tiny bits of what looked like tinsel on the floor, an absence of chocolate that I was looking forward to eating, and the next day, some festive looking poop in the back yard.

Bu got himself into two traumatic and almost fatal situations with his choice of inorganic foods. The year I moved into my current bush retreat, where I now dwell with my beloved partner and three huge hounds, was the year Bu began to fail in his health. It was most likely triggered by a celebratory dinner. Although not traditionally considered Christmas fare, we loved Chinese steamboat and spent hours preparing the ingredients, the sauces and in setting the table to enjoy ourselves. We prepared one side of the steamboat for the meat lovers, my dad and his wife, and the other side for the vegetarians, Tarryn and me.

It was a great night, Bu hanging around the edges, getting tidbits of beef and pork from under the table. After cleaning up that night, we put the remaining vegetable stock in the compost bin but the meat stock we dumped in the backyard sand. Unbeknownst to me, Bu decided to take a midnight snack and siphoned up roughly five hundred grams of sand in order to suck out the meat juices in it. He whined a bit during the night and left some sandy deposits on the couch but being tired from months of renovations and having only moved in a couple of days before, I wasn't paying great attention to his signals for help. It was only when I got up late the next morning and Bu was crying and unable to keep still, I realised he was horribly sick, dehydrated and unhappy with pain. It was Boxing day and most vets were closed and I didn't know who to turn to.

Bu always needed to be muzzled on vet visits as he would start off all trembling and afraid but escalate into an aggressive fighter in the space of a minute and make valiant attempts to ward off any vet attention. My dear friend Emma suggested a vet that was a thirty-minute drive away. I am grateful to this day to the tenacious and remarkable professionals who saved his life, against all odds.

Bu, I was told, was within twenty-four hours of dying. He was so ill, he did not need a muzzle when the vet examined him. Barely able to raise his head, there was no resistance and he just kept still, looking incredibly pained as they diagnosed what was wrong. We got sent home so they could take X-rays, which is when they discovered the sand, he had taken in. They put him on fluids and did their best to save him. The sand he had consumed bound with the mucus in his intestines had proceeded to turn almost solid, nothing could shift it except an operation—they had already tried a flush through. He came home a few days later but within two days, it did not look like he was getting better. In fact, the second day, I woke up to find him lying outside in a puddle of thin but bloody looking water. I took him back to the vet who checked him out and said he had septicaemia. In fact, it was worse than that, the first operation had not had the desired effect and somehow his intestines had become bound to other parts of his body and these required separation.

The combination of already being weakened, taking on another critical operation and septicaemia was very serious and being twelve by then, potentially fatal. The vet gave him an optimistic fifty percent chance of living after the second operation. He worsened within three days and got sent to a twenty-four-hour vet clinic with more equipment and around the clock care but not even twelve hours had passed before we got called to come and get him; he'd ripped out the IV lines and was attacking staff despite strong sedatives.

They said they would give us the medications he needed but I could hear they were sending him home to die. When I arrived there to pick him up, he was severely wasted and could barely stand but wagged his tail madly and although I felt like my world was shattering apart and was quite helpless to properly assist him through, to whatever outcome, I looked him in the eyes, as I had so many years before and promised I would be there, every step of the way.

I went home with six types of medications, something due every two hours, even through the night. It was a time to contemplate the process of letting go of my darling furry boy, of attachment to the physical body and death.

One of my favourite practices from both Zen and Shinto is to contemplate your own death to bring more awareness to the brevity of the life that we have. This seemingly clinical analysis of death is not for the sake of vanquishing fear but rather to appreciate the preciousness of life and in doing so, offer one's self up for compassionate service, time and time again, as a Bosatsu.

Rather than being scared, we should reflect that when fear or death comes, the opportunity to do any reflection on this will be lost. A great Buddhist teacher, Thich Nhat Hanh, had inspired me often with these words,

"When you look at a cloud...and then later the cloud is not there. But, if you look deeply, you can see the cloud in the rain, and that is why it's impossible for a cloud to die. A cloud can become rain, or snow, or ice, but a cloud cannot become nothing. And that is why the notion of death cannot be applied to reality. There is a transformation, there is a continuation, but you cannot say that there is death. Because in your mind, to die, means you suddenly become nothing. From someone, you suddenly become no one...When you can remove these notions, you are free and you have no fear."

The essence of the practice of the contemplation is not actually on the loss of the physical body but on impermanence and the transitory nature of being—death is simply the most obvious state and the most fearsome perhaps, so it's a great illustrator to motivate us to fully live life. To move through life with a sense of impermanence can be liberating even as we feel the despair and allow our minds to drop into a valley of desolation around the process of dying. The process of dying is what is messy, with sometimes great pain, long suffering, the hours by carers required to be attentive and nursing the sick. Being witness to this can be hard, really hard and the joy of seeing a recovery equally as powerful in its intensity of emotion.

Bu fortunately survived to live another three years but each walk after that time seemed somehow more precious and meaningful but had an edge of fear to it—the fear of losing this creature that I was not just devoted to, but partly defined my life. I watched him for signs of ill health like never before and in doing that, I think I stole something of the vitality from the moment.

Impermanence permeates us, I would walk and think. 'Be awake each moment', I would meditate on as I strolled with him. 'Do not squander your or his life with worry', I would chide myself as we played. He picked up on the swirl of thinking and the mild distress with his doggy senses. In leaning against me after a walk or play in the park he seemed to transmit a lesson that I must learn the willingness to embrace uncertainty, live with mystery, and make peace with ambiguity. In this, he was and always will be the greatest teacher I have known.

Two years after he almost died from his potentially deadly sand snack, I had to ask him as he lay once again at the vet, in a cage, fluids being pumped into him—really, just how tasty can a toxic buffet of linoleum, potting mix and silastic actually be?

Bu had raided the shed for these, a curious act, since he was always well fed, had supplements and was medicated when needed and had never taken things off the shed shelves before—not even for a casual chew.

After he was admitted with a grim outcome predicated, I stayed to talk him through being left at the vet for care. I took a photo of Bu just as he looked up at me as I chatted to him and seeing the light in his eyes much diminished, I felt panic at imminent loss arise. All I could do was breathe deeply and sink into a state of awareness of the new lesson that had arrived to awaken me from my complacency.

I had fallen prey once more to placid denial of the changes we were undergoing, both ageing, both getting ill. Bu was peaceful in his cage, being cared for and doted on but I was in the grip of a fierce compassionate shattering of my mind's strong walls to deny death entry. It was a brutal reawakening, these Zen lessons are not kind in such moments but ultimately, they are; what else could have prepared me for the separation from the furry love of my life? He was returned home subdued but within moments of checking out the borders of his home and making sure no rats or possums had taken over his territory, he proceeded to scoff half a roast chicken with his usual appetite.

I was concerned about his advancing arthritis as well as his recent illnesses and replaced a steep, tricky descent, made up of jagged rocks and uneven dirt steps, with a gentle sloping wooden ramp that cost a small fortune. I did this because the back garden is where he spent a lot of time, fossicking around the bushes and chasing butterflies when there were no lizards to hunt or possums to be irritated by in the front gardens. He used the ramp to descend but for some reason, decided it was still easier to climb up the steep hill by going underneath the ramp. Little scamp, he never did take the easy path.

Amazingly, Bu bounced from his inorganic snack raids and proceeded to live life much as before. He would sleep a lot, get loved a lot (on his terms), get multiple treats from whomever he could, be walked slowly and be driven around to his favourite humans' houses. When not busy with snuffling out the bandicoot and kangaroo trails on the property, he spent time just mooching and gazing out into the wondrous world from a spot in the sunshine from the comfort of a soft bed.

Bu's Lesson—The only certainty in life is that we will pass one day, leaving very little trace, so really enjoy each moment, the mooching, the snacking – and above all else, the opportunity to love and be loved.

Chapter Nine
The Eighty-Four-Thousand-Dollar Dog

The medical bills, oh dear me, the medical bills! I thought I'd got an absolute bargain adopting Bu for one hundred and fifty dollars. After much merriment, mayhem, mishaps and inorganic snack misfortunes, I calculated the vet bills, just curious about low bank accounts each month for the decade we had been soulmates. I'd never gotten out insurance, he would not have been eligible with a bunch of issues when I adopted him and it wasn't really a well-known thing. Oh, how I wished I'd known better! Oh dear me, the medical bills!

They went on and on and on…

$150.00	Adoption
$750.00	Anxiety medication, natural remedies and consultations
$150.00	Bowen therapy
$500.00	Behavioural therapy
$1,500.00	Embedded hair on inside eyelid, two eye operations
$750.00	Grass seed in left ear
$750.00	Grass seed in right ear
$1,200.00	Blown eardrum
$1,600.00	Grass seed up nose (each nostril twice)
$2,000.00	Dew claws one (Tarryn's car) and two (the farm)
$3,000.00	Infected anal glands, multiple times, until finally removed
$2,000.00	Benign lump removals, teeth clean and extractions

$2,300.00	Cancerous football sized bump woven between his ribs removed
$1,200.00	Folliculitis in his front paws that eventually warranted an operation where he wore socks for walking for a few months
$1,500.00	Fox poo episode
$34,000.00	Sand snack episode
$15,000.00	Toxic buffet episode
$4,000.00	Several gastro episodes
$6,000.00	Kidney issues
$3,500.00	Arthritis treatments
$2,400.00	Thyroid issue treatment and monitoring

It is stunning to look back on his life and see how full of vet visits it was and how well he weathered the incidents and accidents which regularly featured in his life. The little guy suffered through so much, of his own devising most of the time, but not always. Each time Bu was injured, sick, weakened or suffering, I felt the guidance of Zen's wisdom. Those lessons which had in the past eluded me, I was able to grasp with clarity—sometimes only fleetingly—the rare encounters of true understanding sublime and worth the wait.

Zen teaches me I will die and I will be separated from all that I love and I can take the opportunity with my loved ones as incredibly precious, and not take it for granted. In this contemplation, who my 'loved ones' are, is vast and ever more inclusive. That's why constant contemplation and remembrance of impermanence is important to me because the default seems to lull me into a kind of somnolent, lurching through life. To contemplate the fact of impermanence in order to live an intimate life within each moment, I feel the absolute value in its ephemeral nature.

If I linger over the photographs I have of Bu and enjoy the moment of capture without grasping for the memory, I am intimate with a moment. If I breathe deeply in the remnant scent on his collar and put it away without a sigh, I am blissful and if I can write without regret of the life of Bu and express wisdom tradition lessons in The Book of Bu as I understand them, without thought for an outcome, I am content and willing to let go of the next moments of life into the unknown.

Bu's treatments cost me only money and I believe we need to have some kind of system for measuring how we consume, produce and share as humans. So, to me, it's a form of energy that is easy to exchange and it's not just a necessary evil.

Money can also be divine; it powered the positive activity I wanted to engage in and healing Bu of all the consequences of his style of engaging with the world and his horrendous abuse suffered was that activity. In Buddhism, money is seen as a force for empowerment and it can empower many meaningful things in the world, the finding of a path to Nirvana, rather than wealth. In some humorous synchronicity, it is said the Buddha transmitted some eighty-four thousand teachings. Each precious dollar gave a more precious lesson. Each cent then for me, of eighty-four thousand dollars, was conceived in terms of healing and healthy longevity.

Some acquaintances have summarily dismissed my love of Bu stating in various conversations that this money could have been spent on overseas holidays, the sports cars I so love to drive (I do—I am a Zen petrol head with a history of early model Mustangs, Limited Edition MX5s, Ford Thunderbird V8 Coupes, Honda Legends…), beautiful things to admire in my home, new this and new that. I asked why would I desire any of that over a life that claimed my heart each and every day and taught me to hear Zen lessons like it had never reached my ears or penetrated my heart before?

In offering service when asked, in looking at the value of money differently and in following the path (zigzagging there and admiring all the distractions admittedly) to Nirvana, I am sure I could have and would have offered more.

The devotion to a purity of purpose is in intending to service, I would have sold my belongings, my house and all the other illusions of permanence and security just to be there, breathing in the scent of the undersides of his paws that smelt like malted biscuits, the inside of his ears that smelled like warm honey and being the loving observer of the many mooching moments. Perhaps, these sweet moments in the context of money and letting go of the permanent and my love for Bu are captured in Zen poet Ryokan's verse:

"If your heart is pure,
Then all things in your world are pure.
Abandon this fleeting world,
Abandon yourself;
Then the moon and flowers,
Will guide you along the way."

Although it may be funny to list the medical bills, and speak of what one may give up to devote to another, the deepest lessons for me lay in the illumination of the truth of suffering.

There are eight attitudes or paths Buddhists are taught that they must follow to find liberation and freedom from suffering. Bu's every medical intervention was a chance to pull in the pain and suffering and deeply experience compassion and reflect on the opportunities these presented to be in 'right' state.

For me, these are the 'right' or correct things to do in your life. As a way to reach Nirvana they are right view, right intention, right speech, right action, right livelihood, right effort, right mindfulness and right concentration.

Being in 'right' state when life is flowing and drama-free is easy, curled on the couch listening to Bomber breathe 'BuBu, BuBu, BuBu' was a pleasurable and relaxing evening we often enjoyed.

When life turns to concentrating on witnessing and alleviating sickness and calming agony, it becomes a whole lot harder to be in the 'right' state. When so much is stirred up in me and my mind flashes up horror scenes of loss, I can be restless and sleepless and undecided on what to do, and I question the value of following Zen wisdoms, Buddhist doctrines and Shinto prayers, feeling they are merely a set of distractions. Only in deliberately settling through breath and mediation and cultivating peace do I find my way to being in service with full capacity, competence and capability, and I find myself returned to the Eightfold Path. Peaceful and beautiful paths when metaphorical feet find them again.

In sitting with the ill ease of nursing an injured, sick and ageing Bu, the Eightfold Path becomes obvious, in skipping along the roads of delight and pleasurable distractions, they are obscured. A curious paradox since I would not wish any creature alive suffering and yet, when they are, I have a glimpse of wisdom.

Bu's Lesson—With right intention, your paws can find the right path – travelling on it your open heart might desire to adopt a being with eighty-four thousand needs.

Chapter Ten
The Family Jewels

A long-term friend, George, a man who thinks deeply about life, has an insatiable appetite for knowledge and has a wry sense of humour, got on with Bu remarkably well. He would rile Bu up at dinner parties by grabbing his big block head and smooshing his wrinkled face from side to side and say, "*Who's a bad dog? Who's a bad dog? Bad dog,*" repeatedly, in a friendly tone of voice. All the while, Bu would be thumping his tail furiously in delight, "*I am, I am. That's me,*" until they both would be quite exhausted with the joy of it all.

Bu, of course, took this interaction as a cue to get away with pretty much anything with George. At dinner, he would sneak under the table during the main course and I would suddenly see George go very still and edge his chair back gently and subtly. This would happen as a drooling head, capable of taking down large prey, was dropped firmly in his lap. With the implication that if food was not forthcoming, then teeth may be applied to the family jewels, mere millimetres beneath his lower jaw. I never saw fit to correct what was surely bad behaviour to encourage, in a dog and human alike.

Both routines were repeated for well over a decade of lazy afternoon barbecues, fabulous parties, exotic themed dinners and long, sun drenched lunches. I threw, what turned out to be for many, a memorable house warming party, somewhen in that decade. It was one of those affairs when all the guests from many walks of life gather and mingle easily. Where the afternoon stretches into a night of musicians doing their brilliant thing and night turns into the early hours of the morning. There was simply no way we were going to gather up the forest of half-drunk wine glasses nor the plates littered with bits of uneaten food. We needed bed and sleep and someone to bring us coffee in the morning. All three happened—in that order.

We crawled back out into the garden when the sun was already high in the sky to survey the damage with coffee in hand, only to see poor Bu lying in the grass, with his paws placed over his head—a picture of utter misery. It dawned on us as we tried to coax him up, ready to call the vet to heal his ills, that all the half full glasses were now completely devoid of any wine at all. He had siphoned up every last drop to his detriment; the poor soul had a massive hangover. Bu only came alive when Nannup friend Gary lit up the barbecue and finding left over ingredients, created a fry up that Bu literally salivated over whilst it was cooking and wolfed down with the rest of us.

It cured his hangover and clearly, he had learnt his lesson, as from that time on, he never went near alcohol of any kind and reserved a cute little nose wrinkle for the rare times he detected the bouquet of any kind of wine. Who says dogs have short-lived memories; that one I think, was indelibly printed on his mind.

Generally speaking, most practicing Buddhists do not drink alcohol or at least, do not become inebriated. Zen teaches that when intoxicated, our minds are less powerful, less acute and less responsible. The lack of focus and clarity in this state leaves us partially or entirely shut off to the divine, to the opportunity for stillness and the experience of peacefulness. The same is said of seeking distractions (entertain me boring stooge), not paying attention and in the deliberate seeking of intense experiences.

Of course, as any wise dog knows, we seek sensuality, delicious sustenance and cultivate habits that bring us pleasure; these are all states in which the gateway to knowing of life is there to greet us and is obvious. The more subtle ways of knowing it, life in full colour, are far more elusive and much harder work. Which is why it can be hard to sit down and meditate when a glass of wine seems so much more attractive and within reach.

Bu's Lesson—Meditation does not give you a hangover, try mooching on a cushion for ten minutes or so without any distractions...you'll soon see how amazing the universe is with you in it!

Confident Bu

Art credit to: Nada Orlic

Chapter Eleven
For Later On

Bu would never eat his chewy treats or bones right away after being given them; he would make quite sure that no one was watching and then seek a place to bury them for resurrection. Although he had incredibly sharp and strong claws and could have dug anywhere in any of the gardens or bushland of his various homes, he chose to insert his 'for later ons' into pots, invariably with new plants in them.

Whole raw hides, chicken frames, marrowbones, sweet potato pig ears and snacks of various kinds would be neatly stored in varying sized pots, often with half of these sticking out. He would gently guard his stash by getting up off his outside bed if we happened to walk past them and parade in front of the pot collection like a sergeant major inspecting a platoon that was found wanting. He rarely retrieved his buried treasures and when they began to look dubious or smell, I would dispose of them but I would see him crying softly when he saw me do this. I would normally do this when he was out for a walk to save him the trauma of watching his endeavours go to waste.

He was big on stashing for later on, not just with his food treats but with rotting carcasses found in the bush. A dead rabbit he found lying around the fields of the farm on a walk was brought onto his bed and sucked on. That is until Tarryn managed to take it from him after an amount of disgruntled growling and snarling, right out of those powerful jaws. I thought at the time this was an act of bravery but in fact it was a display of Tarryn's remarkable affiliation with all creatures. I saw him later in the year, chatting to and patting a large bull in a field surrounded by his herd of cows; not something any person can do lightly.

When Tarryn and I were thinking about moving to Tasmania, I sold my house in Fremantle and I lived temporarily in a friend's gorgeous stone cottage in the Perth Hills. It was not ideal to be moving Bu around as he was getting older, but he was pretty strong and adventurous, if a little arthritic and still loved to spend his days outside, sniffing news carried by the wind. Outside the cottage, he had a bed that he laid on in the sun, looking for all the world like he was in deep meditation and contemplating the universal mysteries. He was the Zen master after all.

For a couple of weeks after moving in, I would look out adoringly, call to him and he would sigh, short and sharp, through his nose, look at me with no particular expression and go back to his musings almost instantly.

The weather was getting warmer and so, I would open the window and over time, I noticed a rank smell that got worse and worse. I searched frantically inside the cottage and then all around the cottage and through the gardens going crazy with the stench. Then one day, it rained and Bu was inside by the open fire. When I stood up from patting him and looked out, I noticed that what I thought was a ragged chew toy on his bed on closer inspection, turned out to be a maggot ridden corpse of a shingleback stumpy tail lizard.

I assume that the lizard had strayed in the cottage yard and Bu had unfortunately killed it and not having any pots to bury it in, simply found the most convenient place to guard it—on his bed, under his own body.

Bu cried as I removed the lizard and buried it in the surrounding bushland. He let me know by the angry set of his body and not allowing me to pat him, that the removal of his hard-won treasure was no small matter and he was not happy. Not even a replacement bribe of raw hide could sway him to return to his normal self. He sulked until the next day, when all was forgotten in the joy of a walk around the hills, with fresh post rain scents to sniff and track.

Bu obviously loved hiding his treats until they were rotting, swollen and hideous to smell or touch. The smellier and grosser, the better in his book. One of the rare times he unearthed one of these gave me the scare of my life. He had dragged out a very old raw hide which had unravelled from about six inches to a full metre long. It was not only thick and swollen from the rains, it was stained from being buried in a precisely striped pattern that for all the world looking at it lying on his bed, seemed to be snake. I yelled at Bu to get back away from his bed and I became almost hysterical as he approached the snake and began to play with it, which made it seem alive as it writhed under his paws. My heart was in my throat and I was about to leap in and separate snake from dog when I realised that what he was playing with was already dead, and in fact not a snake when I looked properly, and that I was being silly. That feeling of fear was consuming in that moment; I had feared for him and I was prepared to leap in and take the snake away from him. I wanted to defend him and make his world safe, knowing that I might have to risk my life. Looking back, I consider how my ego is at these times, it swells to gargantuan proportions.

I learnt through my erratic study of Zen that if ego had a physical engine, its fuel would be fear—which far from being something to ignore has its time and place. Throughout my life, I have experienced the extremes of fear, rendering me useless, or as motivation, to make significant and positive change. I learn how to not only overcome my fears, but to become fearless in the practice sometimes called the 'Lion's Roar of Zazen'. The lion is the symbol of self-possessed power, a magnificent beast that has dominion over all he sees as well as the courage, speed and might to attain what is required.

Instead of being nervous or fearful, Zen Buddhists are taught to be eager and open, accepting of all that comes our way. In order to approach life like this, I needed to let go of preconceived notions and be optimistic. This translates to being determined in the face of obstacles; like the lion we can remain calm and still until required, not expending any unnecessary energy on that we cannot change. When we are empty of thoughts, in the calm and still waters of the mind, we act purely in the present moment, without fear, anger, ego or other emotion. We are a gentle force that has tensile strength and integrity. We possess equanimity in being both relaxed and keenly aware of that which surrounds us. Maintaining a relaxed alertness, fear cannot sway you, even in the face of a frightening opponent, a dog almost rabid that his rabbit is being taken away or a bull who is warily watching over his herd of cows and for me, like a snake on Bu's bed.

The wheel of Dharma played out each day strongly in my relationship with Bu. Dharma is a word that has many translations and multiple meanings. Here, I use it in the sense of protecting a life from suffering, harm and problems through developing peace and happiness within the self in order to create the space and offer compassionate service.

It is impossible to sustain the appearance of outward peace if inner peace is not truly present and the way to plant the seeds of peace for flourishing is to grow them through Dharma practice. One Dharma practice that follows on from the Lion's Roar is to let go of cherishing ourselves above all others. Otherwise, we continue our suffering. If we learn to cherish all beings more than ourselves and subsume the ego, we can enjoy the bliss of lightness, maybe even enlightenment.

Dharma is a tricky practice though, but Bu fortunately was my ever-present teacher and I could cherish him and include him as the centre of the universe. I could extend beyond him to people who also needed help and draw them into the centre of the same universe, expanding the circle to encompass more and more, which made me feel not only deeply powerful for doing so but also liberated.

Another way I find liberation is in reflection. Instead of getting angry with someone or some situation, there is a patience that comes from seeing the source of this as a spiritual teacher, a Bosatsu. I can generate a mind of gratitude—thanking the person for the lesson of deceit, of harm or of betrayal—because I now know more of myself in Seishi and the purpose in remaining in it stronger still. We can transform situations of anger and self-pity gradually into a lesson in unconditional love.

It does take both skilful thinking and patience. A person without much patience has a busy and worried mind and is upset by an obstacle or a criticism and reacts. In contrast, when we develop real patience, the mind can be as stable as a mountain and as calm as the deeps of an ocean and we can respond. With such a calm and strong mind, demands placed on us can be met with grace and compassion.

I grew to accept, cherish, love and even welcome the demands of Bu and his special characteristics, like leaving things for later on. How else would I have come to engage more strongly in a deep spiritual practice guided and informed by Zen?

Bu's Lesson—Find someone to teach you patience and how to love others, then the more you can love others, the more you'll find you can love all others regardless.

Chapter Twelve
Bribery and Ducks

When I lived briefly away from the bush, in the northern suburbs of Perth, we found ourselves in a rundown house close to a wetland with lakes and parks that Bu adored running around on his twice daily walks. I engaged the services of an excellent master painter to bring some life to the tired old house I had bought. Frank was a mad Texan with a fondness for sharing funny satirical stories about his rather turbulent life. With a soft spot for all creatures but in particular dogs, he and Bu became the very best of friends. Frank was devoted to his 'Bombardier' and when the smell of paint was too much for me, I would go away and leave Bu in Frank's care (they could not be separated). Frank would stay over, using the spare bed which surprisingly, Bu didn't battle him for.

I am told he would just curl up at the end of the bed, leaving Frank as much space as he needed. I was jealous! When I holed up there to get some space, Bu would only ever leave me a small slice of the real estate of the bed. In the mornings, I would wake to find myself clinging to the edge and cold from the lack of covers that he had somehow gathered beneath him to create a cocoon. He was a great seeker of comfort as you have heard and was persistent as only dogs can be to get it. For those of you who have dogs, I am sure you have the experience of waking up cramped and cold while your dog luxuriates in the vast space and warmth of their creation.

Frank stayed over quite a few times and cared for his Bombadier. At the end of a week away, I returned to find Bu sitting on the front door step, wagging his tail in excitement at my being home. It's so nice to come home to someone who is pleased to see you! I noticed when he got up that he was a bit slow and somehow looked turgid, like an overstuffed cushion. I wondered if he had a problem to be so bloated like that but he didn't seem distressed and in fact looked rather chuffed with himself.

I asked Frank, who had gotten down off his painting ladder, how things had gone. Wiping his brow of sweat with his red handkerchief, he said the Bombardier had been a bit of an issue for him. Apparently, as Frank tried to move his ladder along to paint, Bu would sit right where he needed to land the ladder for the next bit of wall and refused to move until given a treat. I knew I hadn't left that many healthy treats around so I asked, "*What did you give him?*"

"*Well,*" Frank said, "*The only thing I could find were those Dutch almond rounds you guys like so much.*" A picture was sketched in my mind and then it got filled in. "*And you know he would only go once I fed him the whole thing.*" Bu had managed to swindle his way through not one but two packets of Dutch almond rounds and in a twenty-six-kilo dog, that is an incredible amount of not just food but entirely the wrong kind of food.

Poor Bu was put on a diet which he did not appear to enjoy one little bit that was especially formulated to shed weight fast but maintain nutrition. His potted treats had been removed to avoid the temptation of him breaking the regime of calorie restriction and I sure enough watched him checking all the pots for the sequestered goodies.

During the time of his diet, a couple of ducks landed in the pool and chose to stay for a few days. A great game of the ducks gently paddling out of his way each time he would run to the side they were drifting to ensued. Round and round the pool he would run, barking and leaping, always to find the feathered friends gliding away serenely. He would give up for a couple of hours and watch them with a degree of resignation until the fire in his belly for hunting would light up and round and round the pool he would go, taking care not to slither in.

Oh! How he hated water! If he stepped in a puddle, he would shake his paws out one by one like a cat. He would hunker down in a sudden downpour of rain and get out of it as quickly as possible. If I tried to take him out in wet weather when I wanted to go for a walk, he would sit down, attached to his lead and suddenly, somehow make himself as heavy as a hippopotamus and as recalcitrant, refusing to move a paw out of the door. He could not and would not be moved, he put me in mind of the Wicked Witch of the West in The Wizard of Oz who melts into a puddle when water touches her, 'Melting, Melting, Melting…' In any case, all that chasing around after the ducks helped his weight loss and he was soon back to looking his energetic, muscly, sleek and glossy red brindle best.

With an acquired a taste for pursuing ducks, he would chase them at any opportunity. Not that he had many, he was very well supervised for the most part but as an adept escape artist, he did slip his collar at a friend's birthday party a few years later and headed straight down to their half-filled dam where he chased the ducks round and round the enormous perimeter. His high-pitched yapping carried across the farmlands and he sounded demented in his over eager pursuit of his elusive quarry. It took me risking a broken ankle or worse to catch him and pull him up the steep embankments, away from duck dinner.

He rested well that evening as did I; it was more excitement than both of us could handle for one day. We drove home, Bu in the front seat stretched across the middle console so his head could rest in my lap, doing his happy breathing 'BuBu, BuBu, BuBu'. I loved him so much. I recall my heart squeezing tight in the ecstasy of connecting with his breath. His fur spiking though my best jeans and his round brown eyes looking up and letting me know it had been a good day for a dog. There had been ducks to chase, pats from friends, tidbits of birthday cake, a ride in the car, a walk in the bush and so it had been a very good day indeed, a day of no particular purpose or meaning.

'The Meaning of Happiness' and 'The Wisdom of Insecurity' are two books by Alan Watts that struck me as encapsulating the themes that life has no intrinsic meaning, no matter how hard the mind might try to grasp this. It is best then not to be attached to events, as these are merely a thing happening in a period of time, and we are emerging out of the world and our minds, not into them and expressions of that which we are interacting with and experiencing. Zen is about reclaiming and expanding in the ever-present moment, embracing the transitory and letting go of the illusion of future. To Zen practitioners, holding onto past, present or future points in time is absurd and pointless as there is no meaning except perhaps, in that infinitesimal moment.

The trick, of course, to a good day is to emulate a dog's utter devotion to living in the moment. Savouring the sweetness of bribery cake, blissing out under a stroking hand that may be removed at any time, enjoying the rush of the air past the head as it sticks out a car window, taking in all the smells at once in the bush until there is giddiness, feeling the thrill of the chase of ducks, and finding a kind lap to lay one heavy head, maybe not forever but just for the length of the journey.

Bu had, at some point in his life, before being adopted by me, been near starved and had an endearing habit of coming up to me after each meal and nudging my legs and looking into my eyes as if to say, 'Thank you', just for the length of two heart beats.

It was not just a seeking for more, because there wasn't any more and not simply trying cute looks on me as a form of bribery to get me to reach for the treat box in the pantry. It was simply pure gratitude.

Bu's Lesson—There is only this moment so find a kind lap to rest your head on (preferably one that does not mind your drool) and feel grateful for each and everything that comes your way—good or bad. The balance will be towards the good, I guarantee it with both paws.

Chapter Thirteen
The Human Whisperer

I thought of myself from time to time as a dog whisperer, at least for Bu, which is a term that refers to someone who has a natural or mysterious affinity with the canine species. Bu was a human whisperer and was the keeper of all my secrets. His ears were filled with the comings and goings of my daily life and the most significant moments of it too. I wondered if he ever thought about the content of my words or if he felt the weight of the disclosures of heartbreak, of stories of failing, of undiminished longing for something more in my life to lift my spirits high. He seemed to accept it all very serenely and like a wise sage, offered no comments of advice. He simply, as an advanced being, knew that all answers were within me and he was merely the sounding board.

Somehow, I always felt better after I had talked with Bu and almost blissfully drowned in the liquid pools of his eyes. I was silently whispered, cured of ill feeling, lifted from not succeeding where I thought I should, and most definitely felt calmer and keener in spirit.

I wasn't the only human he whispered to. As he aged, he required a midday medication for his arthritis and a walk to help loosen him up and allow him to be comfortable for the afternoon of snoozing ahead. As I worked fairly long hours, I employed carers to do this for me and we were truly blessed by those compassionate helping hands that showed up in service. His first extraordinary carer was an American veterinary technician, Paula. She lived just two streets away, had a pack of rescued animals, both canine and feline. She was familiar with dealing with traumatised animals and Bu's peculiarities were of no concern to her. She was the only person besides me that was able to touch his paws without getting a warning growl. Paula was a fearsomely strong woman in all aspects of her being but with a peculiar softness deep inside that showed in her smile. She was so in tune with who Bu was in the world, she demonstrated no fear at his bared teeth. She could brush his feet and soon after being able to do that, massage his hips in preparation for a walk.

Paula grew to love Bu in a very short amount of time, admiring his tenacity of spirit and ability to simply be exactly as he needed in each moment. They went for a wander every day through the bush or around the suburb. On the days he did not feel well, which was obvious—his eyes changed from perfect round circles to tight painful wedge shapes—she would sit with him and tell him about her problems. Like me, Paula would feel better for sharing the load and observed how good he was at listening. I felt relaxed knowing he was in such expert hands and so, when Paula moved away to live in the South West amongst the tall karri trees, adopt more crazy critters of equine, ovine and caprine nature, I reached out once more to find someone to help and a Hungarian angel appeared.

Rose not only looked after Bu's medications and midday walks, she also did some housekeeping and all of a sudden, I had spare time each day and found more space to meditate, enjoy yoga and an evening walk—all of which Bu would be in attendance for. The sound of 'BuBu, BuBu, BuBu', rhythmically accompanying whatever I was doing was a delight to me. He could get quite loud in his snoring when deeply contemplating his furry navel (he swears he wasn't sleeping) and it was a fantastic way to remain mindful during both meditation and yoga. The sound was not a thing to like or dislike, but something simple I could pay attention to and then let go repeatedly as it entered and left my consciousness.

Rose formed a loving bond with my dog and she would sometimes stay and join me for dinner and Bu would rest at her feet, looking up adoringly at his carer—silently protecting the secrets she had shared with him during their day together. I loved meeting her, coming and going from the house, and experiencing the strength of her character and hearing her robust laugh, in contrast to her exotic and stunning looks.

We were both aware of the huge gap she would leave as she left to travel around Australia and then return to Hungary for a long stay. Bu looked genuinely sad when she said goodbye to us, her energy radiant and beautiful as she looked towards her adventures.

I know Bu was tuned into people in a most special way when he loved them. He knew twenty minutes exactly before I was due home—no matter what time I was actually going to arrive, that I was inbound. This was corroborated by my husband at the time Justin (who Bu did not like that much and bit on the finger), my former partner Tarryn (who Bu loved and looked to for direction), Frank the mad Texan and my dad who looked after him from time to time if I was away for any reason.

Bu's next carer lived locally and tended to him almost right to the last day of his life. Stacy had responded to my advertisement for 'Seeking Loving At-Home Daily Care' and over the next two years, she gave me the most incredible gift of her time and energy.

Stacy kept my beloved Bu occupied and walked, my house clean and tidy and my garden looking amazing. She radiated compassion for us both and she turned up each and every day, serene and sweet, carefully checking in on Bu and how he was feeling, and how I was in relation to Bu's health and wellbeing.

I have kept the book of notes we wrote to each other about what needed doing. I trusted Stacy so much that she took Bu to the vet when needed. As you know, Bu did not like the vet and turned into a 'wee timeous beastie', trembling all over at the mere thought of something happening to him and getting aggressive in the process. Which it did, often, to the tune of eighty-four thousand dollars' worth!

At each visit, Stacy handled him calmly and instead of hearing that he had chewed the finger of the vet, which he did once, *despite me warning them no less than three times that he was a dangerous dog and needed muzzling,* he was reported to be well behaved after being muzzled and easily bribed with treats afterwards. Compassion comes in so many forms and the service to us both, from Paula, Rose and Stacey was exemplary of the Bosatsu whose presence alleviates suffering by taking up the load in grace.

Compassion, as I have grown to understand it, is at the very essence of a mindful life. It is essential for developing a spirit that listens to what is asked of it unspoken, to feel the subtle changes that need responding to with grace, and to observe where help can be given without intrusion. Compassion is taught as a central tenet of most Buddhism schools and is represented often as three intertwining treasures of Buddha as a yellow jewel, Hō or Dharma as a blue jewel, and Sōgyō or Sangha as a red jewel.

The significance of the jewels lies in a commitment to Buddha within, an aim to become someone who sees the nature of reality absolutely clearly, just as it is, and lives fully and naturally in accordance with that vision. Dharma refers to both the teachings of Buddha but also the mind that discerns and reflects unmediated truth, navigating the complexities of life simply and finding ways to transform oneself—such as offering service and expressing compassion. Sōgyō encompasses the way in which we approach life, how we learn and who we learn from. Bosatsu embody this and appear to us with the skills to assist and alleviate and help. Sangha can also be translated as community, a place to be in refuge or retreat.

Buddha once said that *Kalyana Mitrata*, 'friendship with that what is beautiful', is the whole of spiritual life. The paradox of this is that we must love and embrace even those things we are afraid of and seem heinous, as these are reflections of human nature. Bu was beautiful to me and we were great friends and he was part of a wider community of compassionate beings who loved him and whom he loved back.

Bu's carers, every one of them, presented minds that were motivated by cherishing other living beings and clearly wished to release them from any suffering. Sometimes, I notice in myself, that out of selfish intention, I can wish for another person to be free from suffering because I am deeply attached. I would say that I am a failed student of Buddhism as this is a lesson I learn and relearn daily about non-attachment.

Non-attachment is taught as practice 'antidote' to the attachment issues described in the Second Noble Truth. For if my grasping and sadness is a result of finding my life unsatisfactory, it follows that non-attachment is a condition conducive to have satisfaction in life. The philosophical advice to detach or unattach from family, beloveds and experiences, is not to cause distance or separation but rather unwind the illusion that we have a 'self' that exists separately and independently from other people and phenomenon. From there, we may suddenly recognise that there is no need to detach or unattach, because we have always been interconnected with all things at all times. Buddhism teaches that we are inherently joyful and that it is really the mind that surrenders and relinquishes its misguided habits and preconceptions that allows us to experience the Buddha within.

If my dear friend Bu was ill or depressed, I wanted him to recover quickly, yes, so he was well but also so that I could enjoy his company again. I missed walking him in the bush and the peace that brought to me too but perhaps this wish was basically self-centred and not true compassion? True compassion is necessarily based on cherishing others, not the selfish ego and it's grabbing desires and attachments.

Although I already have some degree of compassion, by the grace of Bu's friendship and love, at present it still seems biased and limited to me. When my family and new furry friends are suffering, I develop compassion for them easily, but I find it far more difficult to feel sympathy for people I find unpleasant and loud—those who are not kind and gentle—those I would not desire to be in company with, or for strangers. If I genuinely want to realise my potential of flowering the Buddha seed to full enlightenment, an ever-widening pulse of expansion of compassion until it embraces all living beings without exception, is the key. I visualise this naively as a huge golden key that is magical and heals all that it unlocks.

I feel like when I lost Bu, my golden key dissolved in my hands and grasping at its absence, I then lost my way to the door that was already opened and welcoming. I sat in miserable silence in the middle of my life's path that I no longer recognised as one I had chosen and was creating moment to moment. Deep down, I was sure that a Bosatsu would come along and sure enough, he did; in surfacing my memories of Bu, I woke up one morning and discovered it had been but a moment of suffering caused by my own illusions.

The Buddha seed is ripening and partly opened. I can feel it unfurling in the recollections of Bu in these stories and his extraordinary influence on my life. His human was whispered well and truly as he slowly denuded from my life greed, hatred and delusion or what are known as the three poisons. These poisons are made more swift and potent in their action by that special, sneaky, tasty and spicy ingredient of self-delusion.

The good news for me is that the antidotes to these were gifted by Bu and I held in my hands the four awakened qualities of loving kindness, compassion, joy in the joy of others, and equanimity. These were vastly expanded by aiming for selfless offering. I have a very strong feeling that Bu and I whispered each other from the moment we met and were each other's unwitting salvation.

Bu's Lesson—Surrender and be dog whispered or be your own whisperer, feeling the power and beauty that abides within for true transformation (as refuge savage to loving companion).

Chapter Fourteen
Jesus Lady!

Throughout Bu's life, he was much praised for his stoic bearing and his handsome appearance and funny antics but to some, he could sound and appear quite fierce. When we lived in the dry and cookie cutter housing in the north of Perth, Bu was sequestered away behind a tall brick wall and we hardly saw our neighbours. One day I bumped into them going out for a morning walk, and they exclaimed, "*Jesus, Lady!*", being quite surprised at how small Bu really was compared with his terrifying barking and growling—"*We thought you had gigantic Rottweiler back there—we always go around the long way just in case.*"

He did have a fearsome bark on him, but my Fremantle neighbours, who grew to be great friends and still are today, knew his softer side. After living in a funky limestone cottage behind them for a couple of years, John, a tall man who delighted in telling amazing and often hilarious stories about all his incidents and injuries, mentioned something to me. They heard me 'clip clopping' down the driveway past their bedroom window each morning. I was alarmed that I was waking them up but that wasn't what disturbed them. It was Bu sitting at the gates crying and whining for me to come back. I never knew that he pined for me to be with him until I was told. I had assumed this fierce little creature would be content in his own company. Fortunately, Bu was rarely left alone as he had a succession of people who were with him. First Justin for five years who worked from home, then Tarryn who often stayed over for a few days while he studied the mysteries of the universe on and off for six years. Then of course his paid and well whispered carers. I wondered if he felt lonely and if each day, his heart broke when he was alone. I know he slept a lot, like most dogs do when nothing else is going on, and time passed in doggy dreams.

The experience of being alone can be an intense one, at least for me at times, acutely alone; despite the abundance of friends and family and the comfort offered by those also mourning the loss of him. In the first year of feeling alone, the sense of desolation inspired a warmth towards humans and creatures like I had never quite experienced. It fairly pulsated in my centre, even as I grieved, nursing a broken heart and trying to resolve its sheer discomfort. The more I felt the absence of Bu and the desolation deepened, the more warmth welled up and was available to give and offer compassion when asked for. While still seeking resolution from the discomfort, in quick succession, I adopted the three beautiful refuge dogs Diesel, Stella and Jessie with my darling partner – nicknamed Fuzzy Bear (because he has a hairy chest and stomach and tells funny jokes—wokka wokka). Incredibly, I found myself more alone than before! It was not 'dog' I missed at all, it was Bu—my furry soulmate and the hardest lesson of all began for me. I felt I deserved resolution, yet I was suffering from the seeking of it.

Tripling the amount of canine companionship couldn't resolve Bu's absence. Slowly, it came to me that an open state of mind that can relax into the paradox and ambiguity of life and death would provide solace. It would be something more than attempting a pale victory over the fear of loneliness and loss. I wanted instead, to stay in the knowledge that we are fundamentally alone and there is nothing anywhere to hold on to, not a job or a house, not a beloved lover and not a Zen dog named Bu.

In meditation, and I am a slow learner and need this lesson over and over again, I discover an unfabricated state of being in cultivating that state of being alone. I am disintegrated into my essential parts and drift a disbursed, insignificant and a lone soul who is nothing extraordinary after all.

When those closest to me saw me sell my much-loved MX5, one of them said, "*Jesus, Lady! You must have loved that dog.*" You see, I had sold my prized possession to pay for the vet bills I had accrued. It surely stung seeing someone else driving my car away but then, it didn't really define who I was. In letting this go, not just for practical purposes, I was practising detachment; knowing that it would one day have been stolen, broken, gotten antiquated and discarded for scrap or sold. I was very happy going out for a drive with Bu sitting pillion, his tiny ears flapping and snout sniffing the winds. I got high on driving fast and got a kick out of people waving to us when the top was down—Bu and me stepping out in sports car style. I loved the joy rides, but happiness was not in the car itself, it was not in what I owned but rather in the drive itself with my companion pooch and how it nourished the relationship.

Bu was not my only passenger. Sometimes Tarryn was. On occasion, Taryn would drive, often at high speeds. Being skilful, I had complete trust in his handling of the car and laughing, I would throw my head back and look up at the sky and the tree canopy as we raced by and think—how does it get better than this moment? It never, I knew, would, because there was only ever that love arising in that moment, and the next one, and the next one... I wanted less and less after I sold that car. Zen taught me that when you want less, have less, and then need less, gathering energy to make a change is a much smaller effort and we are liberated to pursue that which we love to do. Something I love to do is explore the delicious space of just being, not doing and being alone liberates the energy to do this.

Bu loved a few toys, his possessions, and he knew what was his and what was not. He had a basket full of distractions and things to chew on. He would often be seen after his evening meal, furry little butt in the air, diving into it and retrieving the special toy of the day. He'd throw it up in the air making it roll away from him so he could chase after it, or to encourage me to take one end of it for a game of tug-of-war. For a dog of only twenty-six kilograms, he was remarkably strong and would wear me out quickly, so he would drop to his haunches and play the 'bitey bitey' game which consisted of him chewing my forearm in a gentle wrestling motion which left a disgusting amount of spit on me—not to mention a few indents when enthusiasm overtook care.

Bu had tiny front teeth, two huge long canines (minus one from the thunderstorm episode) and crumbling molars, worn down from stress grinding and a bad diet early in life. It never seemed to faze him and he compensated with jaws that were probably in the distant past, made for bringing down huge prey. Bu had a really big head and John, my Fremantle neighbour, would often look at him as we went out for a walk and comment mildly, "*Jesus, Lady! Your dog has a great big head.*" He always gave Bu a pat and mushed his wrinkles around much to Bu's delight. He teased him with, "*who's a stupid dog then, who's a big-headed boy then?*"—leaving Bu wriggling in delight and amped up for a stroll when he could go and find those sacred scat treasures, he adored so much in the remnant Fremantle bushland.

Coming out of the bushland once, I recall we came upon a group of people sitting around, drinking. They looked like they had been there all afternoon and Bu being off the lead went to say hello. He stalked quietly up behind one man and nuzzled into his neck without warning, making the man jump up and turn quickly, and looking with concern at me, he yelled, "*Jesus, lady! What the hell kind of dog is that?!*"

"*A good one,*" I replied, "*he just wants to be patted.*" He sat back down heavily, realising he was quite safe and Bu submitted to some petting and cooing from him, leaving a trail of slobber on him as a token of his momentary affection.

Bu would arrive home from his walks pretty tired most times but if he had extra energy, he dived for a toy in his basket and played with it until truly exhausted and then slept. I calculated Bu went through about twenty dollars of chew toys a week and I decided if I paid more, then I would get hardier toys. I fetched up at pet store, with credit card in hand, prepared to pay any price. The helpful young man said, "*Well, we just got these in, they are used to train tigers—they are guaranteed.*" Joyful, I took one home, thinking of all the money I had saved. Bu stuck his nose in the bag when I got home and took the new toy out. He snorted with excitement, bypassed the sweet potato pig ears and took it into the lounge room to chew on. I went and made a cup of tea, feeling very pleased with myself and happy for Bu that he had a toy for a lifetime.

About ten minutes later, with tea in hand, I walked into the lounge to find the toy demolished, the floor littered with small pieces of it. I set my tea down, gathered up Bu and the pieces of the toy and drove back to the pet store. The helpful young man was astounded when I showed him the evidence. "*Not dog proof, let alone tiger proof I suspect*," I said as I handed over the pieces, teeth marks in view on each fragment.

"*Jesus, Lady! What kind of dog do you own*?!" he asked. I showed him Bu, the pint sized Nannup tiger, sitting proudly in the car as if he owned the store and all the chew toys in it. The guy shook his head, declining to get his hand near Bu's head who was wagging his stumpy tail madly for a pat and walked back inside with me. I got my money back and continued to pay about twenty dollars a week in chew toys.

I thought to myself, *Jesus, Lady! Don't sweat it—it's not even a drop in the ocean compared to the vast fortune spent on medical bills!*

Bu's Lesson—Never underestimate the amount of love and liberation one small determined creature can chew through – and foster in your life.

Chapter Fifteen
Yum – First Blood

In a rare car cleaning mood, I stepped out under the shady branches of the hundred-year-old oak tree in my Perth Hills property. Bucket of warm water and vacuum cleaner in hand I marched to the car, with Bu trotting faithfully by my side. He was keen to investigate the unusual occurrence of the situation and see if any rats or frogs needed chasing off in the near vicinity and prevent interruption to my duties.

I washed and polished the outside of the car, I wiped all the windows sparkling clean, I rust proofed small areas of exposed paint and thought I should tackle the inside of the car and remove the dog hair which decorated the entire back seat and adorned the front seats too. Bu had barbed hair and some hair required nothing short of tweezers to extricate it, one by one, from the fabric of the seats. I vacuumed the inside roof, the floor protectors, the front of the seats—which looked like they were sprinkled in red and orange coconut—and then, still having energy and having come so far in the process, decided to vacuum under the seats.

I pushed the seats forward by the side lever on the left and balanced myself on the right of the rail. I began to vacuum and, in my enthusiasm and carelessness, managed to jam the nozzle of the hose under the lever and in doing so with my finger on the rail, the seat rolled over my right hand, ripping off two nails and separating a good chunk of skin under them as well from the muscles beneath, I was screaming in pain. I recall it being almost unmanageable and sickening but I needed to remove the chair off my fingers and had to use the nozzle in reverse to do so—taking off more nail and skin. My fingers were a mess and blood had sprayed everywhere all over the car seats and on the newly cleaned roof.

Once freed, I realised I was bleeding very heavily and going into shock as I was quite woozy. I headed for the kitchen and noticed Bu trailing behind me, stopping and licking up the blood drops along the path to the back door. He showed every sign of happiness and even stretched himself up on the kitchen benches to see why I was washing all that good stuff off and if there was more to come for him! My neighbour had heard me and came over to see if I was in need of help. He called an ambulance as the damage to my fingers was quite bad. I turned to secure Bu in the house but all I could see was his brindled behind sticking out of the car door and getting closer, I heard a disturbing slurping noise.

I peeked my head in to see what he was doing and grabbed his collar to get him inside—he had a hatred of uniforms and I assumed he would attack the ambulance officers on sight. He was oblivious to my presence as he licked and sucked at the seat covers to extract my blood. He then proceeded to hop up on the back seat to get at the sprayed blood drops on the roof. It was slightly horrifying to see his savage animal instincts had outweighed being my adoring companion.

The ambulance officers decided I needed emergency attendance and I got a smooth ride to the hospital, was seen to very quickly and operated on the same night to put my fingers back together. I slept at my mother's house—it was the first and last night Bu ever spent without human company. My neighbour did try to look in and feed him, but Bu was reportedly unhappy with him coming near. A few growls showed him off, his animal instincts at work to protect his territory.

Zen, Shinto and Huna all demanded of this student to show equal care and compassion for each and every creature in the universe whether they can become enlightened or not. The destruction of any creature represents a lack of compassion and is considered an expression of greed. The Buddha's teachings are clearly against any form of cruelty to any living being.

I have been over my lifetime, a strict vegan, a vegetarian, a pescatarian, a meat eater and revolved through those doors a couple of times and landed back to the state of vegetarian—the vegan path being too strict for the way I choose to live my life. I opted to follow the vegetarian path as a statement of protest for the way we treat animals on the planet as well as respecting the Buddha's teachings. It's not difficult for me, I enjoy the feeling of lightness the diet brings and the mindfulness it takes to live this way.

The consumption of meat is a contentious one in society generally, people's views and practices around it vary, even amongst the different types of Buddhists. A vegetarian lifestyle is encouraged in Zen practitioners as it follows the example of the Buddha who was recorded in some of the Sutras as speaking out forcefully against meat consumption and was unequivocally in favour of vegetarianism. The eating of the flesh of fellow sentient beings feels incompatible with the compassion that a Bosatsu should strive to cultivate and encourage in others. The Sutras' meanings and intentions have been questioned, just as the Buddha predicated they would be, long after his death. I think this teaching goes further than whether we consume meat or not or cause harm to an animal or not.

Today, we farm and destroy animals and in the process of growing them and selling them as if they are unfeeling commodities, often deprive them of their natural rights. We are already beginning to pay the price for these selfish and cruel acts. The vast scale of farming of animals in the manner we do, is threatening and has permanently damaged many areas of our environment. If we do not take disciplined measures for the care and survival of other creatures, meat producing or not, our own existence on this Earth may not be guaranteed. If you want an interesting read on the subject, try '*The Face on Your Plate*' by Jeffery Masson who takes not a Zen view of eating meat but rather a psychoanalytical view on how food effects a human's moral self, our health and our planet.

The Zen ethic is concerned with the principles and practices that assist us to act in ways that help rather than harm. Lay Zen Buddhists try to live by Bosatsu Kai or five precepts and act in accordance with these. These are: not harming any living thing, not taking what is not given, misconduct of all forms, lying, and consuming intoxicants. To live is to take action which can have either harmful or beneficial consequences. The five precepts are not rules or commandments, but guides by which to practice with intelligence and sensitivity. So, rather than speaking of actions being right or wrong, Zen speaks of them being discerning.

It is a paradox then that I buy meat and make food for my three dogs that comes from farmed animals. They do have vegetarian meals but are clearly not impressed by them and toy listlessly with the contents of their bowl—they are designed to be carnivores and I accept that and having promised to take care of them, I give what is needed. Although in doing so, I am going against my own ethics and morals concerning animal welfare and environmental care. I know I am indirectly harming some creatures to satisfy the appetites of others.

There is definitely blood on my hands in the purchase, in the mouths of my three beloved pooches from the eating, on the butchers' bodies as they slaughter animals and on the counters of the stores that meet demand by supply of pet food. It's an endless chain of production and consumption and the path to break it, for me at least, is to have no more pets when Diesel, Stella and Jessie have lived their happy, furry lives, feeling safe and loved. Each of them has a harrowing beginning with stories of brave rescues and they each deserve the abundance of health they enjoy, however that is given and received. Perhaps all the contemplation on ethical and moral issues of keeping and feeding meat-eating dogs is simply not accepting the natural order of things that unfolds in the world of prey and predator.

When I bought the house I now live in and opened the front door to Bu for the very first time, he ran around excitedly from room to room, sniffing at all the brand-new smells and wagging his tail in happy delight until he came to the room that would later be used for a study. At the door, he halted and keened, his tail stiffening directly out behind him. He proceeded forwards very slowly with his nose to the carpet until he hit a reddish-brown stain. Laying down in front of it, he sniffed carefully around the edges of the suspicious looking patch and to my mild horror, began to suck at the carpet sounding like a little kid draining the dregs of a milkshake. I guessed that he had his second taste of blood that day and it took removing the carpet for him to abandon hope that any more could be extracted from it.

I tried to feed Bu a vegan diet for a few weeks, but it was not the natural order of things for him. He simply did not thrive on it, although he ate what was given to him. He was morose in his demeanour and lacking in that specialised 'Komatsu' bulldozer quality, with which he conducted his life. Bu was the ultimate carnivore (supplemented by poo, let's not forget) and wanted his meat—the bloodier, the better—mine for preference!

Bu's Lesson—Live gently while listening to and living by your instincts, they will guide you to be in the perfect place, at the right time – the treasures you seek will be there.

Chapter Sixteen
The Chase

Most dogs I have had the pleasure to know will, when taken for a walk, sniff along the borders of the path they are walking on with perhaps the occasional foray to either side to sniff at the bushes, snuffle at the undergrowth and collect the news of the neighbourhood from the trees and lamp posts. Bu was disinterested in this, except of course, if there was a small winter stream running down the path and then, he would poke his head as close to the water as he dared and seek out those pesky honking, hooting, moaning frogs.

He preferred to go much farther afield and roam in search of what Tarryn fondly called 'airmail'. He would frisk around a park's confines or along a tree-laden path in search of dropped and broken branches. Nothing pleased him more than seeking the freshest, hot-off-the-last-dog news and run around in excited little circles, sniffing in the gossip. He *always* left a reply, being a polite dog, sometimes for emphasis, a couple of lengthy and possibly poignant replies.

I am not sure what the attraction was to fallen branches, logs and broken bits of debris from human existence. I feel it was the delight in the path less travelled and the spirit of the explorer in him, craving to be satisfied. The finds made him stop in the middle of a full tilt boogie run, pulling up where his nose found 'x' that marked the spot while the rest of his body swung around into alignment.

He loved the ruins at a particular park down Fremantle way; we fairly wore a groove around the lakes, walking every day through the scented and wooded grounds of a heritage house and pottered about the limestone ruins studded with small native flowers around and near the bush. It is a magical place at any time of year, the house a little way from a wetland, well preserved and flourishing, with all kinds of bird life—musk ducks, herons, and even blue billed ducks. There were butterflies of all kinds and dragonflies darting in and out of the reeds. Bu never chased any of them, I suppose it would have meant getting his feet wet, so he only gave chase on dry land or up palm trees if he could climb them!

John, the man of hilarious stories and Natalia, a woman of great intellect, and someone who inquires continuously about the world and is fascinated by everything, owned a cat. Missy Pulu took great moggy delight in perching up on the posts of my gate, peering down, and well out of the reach of Bu, hurling cat taunts at him. How he would rage! No matter how high he stretched, jumped and tried to climb the posts, he could not get that naughty cat. Giving up, he would walk away with his head drooping and go inside to wait for the thrill of a proper chase.

Pining for the chase is a daily grief that I, and those I speak to, often experience as humans. It has a distinct texture of the weavings of a fabric which are the 'not enough' weft threads, 'grasping at security' warp threads and the 'separated from' patterns that emerge. This delusional fabric is lustrous and somehow intoxicating to me. I sometimes act as if the fabric is real, spending a long time creating it and showing it to others to admire, who like ourselves, can't actually see it but assure us it's real and incredible. It costs a bunch of dollars and stress to achieve this great invisible creation. For me, it conjures up the children's story of '*The Emperor's New Clothes*'. We endlessly add to its size and complexity and instead of gaining wisdom and health, we are left with yet more to hold and protect. Which worries and distresses us. It plagues the mind and causes disturbance—so very far from Zazen.

Zazen can be translated as seated meditation, but this is possibly a misrepresentation of the actual practice. It focuses more on the holistic integration of body-mind-breath, allowing the mind to exist without giving it any pre-eminence, allowing the body to sit upright, without giving it too much focus either, finding gravity to weigh it down and cause stillness, while using the breath to foster the cessation of thoughts.

Cultivating quiet control in Zazen may take moments or may take years, but this balanced state is not something to chase. The lesson of Zazen is to shed the layers of delusion by practice and even then, this should not be a goal as such.

What if we were more like Bu and in Zazen? I have wondered then and since. He was released of himself in each moment, he was present in the chase, be it pursuit of news airmail or lamp post or marauding possums or testy cats and then not in the chase. He would be resting and observing each moment passing through him (he swears he was *not* snoozing and was in Zazen) and so, not at all confused by a grasping, grabbing and greedy self, constructing a near impenetrable fabric of self-delusion. His fabric never anything but a weightless cloak that could be discarded instantly when not needed and it was permeable always to allow the light of life to flow in and out and illuminate his doggy dreams.

Reflecting on this, I witness the same in Diesel, Stella and Jessie—so perhaps, this is the ground state canine being?

Sometimes, no matter how Zen a dog you are though, there is no way to stop the chase, the canine instinct is activated. I was taken aback more than once to see my brindle baby on the run, pellets of poo being fired out of his backside as if he were a husky reined to a sled, pounding the snow in the wildness. He was in pursuit of the smells beckoning to him, news laden or with the scent of kangaroo tantalising just out of reach, there was just no time to stop the chase!

Bu's Lesson—Pursue that which makes you happy, chase until you have run out of puff and no more. Satisfaction is in the pursuit itself and is fleeting...if you catch the object of your desire—what then?

Chapter Seventeen
It's My Bed!

Bu thought the spare room bed was his property and no matter where I lived and what room it was put in, it was his bed. Guests were not welcome, and he made sure everyone knew it. He behaved exactly like he would when someone dared try taking a marrowbone, a rotting rabbit head or anything else away from him that he did not wish to part with (cue un-Zen like sounds of growl, snarl, growl).

My aunt and cousin, both dog lovers, came to stay at my house once from where they lived in Melbourne, an eastern state of Australia. At bedtime, the cute Bu who had been laying on the couch having his tummy tickled and ears rubbed, would turn back into Bomber the refuge savage.

Bu would not get off the spare bed, where he would usually spend his days and sometimes nights with me when I needed a quiet space to finish my sleep. He did not want to give that bed up. He growled and snarled convincing my aunt as she tried to get into bed that his teeth were sharp as razor blades and his claws ready to gouge flesh! He let her know in no uncertain terms that this was *his* territory. My aunt called to my cousin, who was met with this similar reaction when trying to manhandle him off the bed, he tried to bite after the growling and snarling routine did not get him anywhere.

I was called in to help and the furry love of my life suddenly turned reasonable and loving. Bu wagged his stumpy tail, '*thump, thump-thump*', immediately hopped down off the bed and looked backwards to give a long loud sigh that seemed to say, "*You really just had to ask!*"

It was not the first time he had bailed someone up in the same circumstances; he pulled the same tactics on Justin's sister, Annabelle, in the house I had in the Perth Hills. I have the suspicion, now that I am writing about these episodes, that he actually enjoyed lying in wait to surprise and terrify the unsuspecting. He'd lull them into a false sense of security, having played the adorable puppy act on the couch before retiring time for a couple of hours. As they were brushing their teeth, he would go and wait in the dark, curled up exactly in the middle of bed. He would look up, sleepy and innocent, so I was told, and the moment they tried to get under the covers, off he would go. Frightening indeed he was, especially if you knew he was known to bite and could demolish toys made for tigers in under five minutes.

He even tried it with me once, I was surprised and yelled really quite loudly, threatening him with all sorts of things—withdrawal of walks, treats, cuddles as I was so scared at this snarling apparition. I backed out of the room to think how I should be in the situation as opposed to how I was reacting. *What was in me that was appearing in him in this moment?* I asked myself.

This is one of the first questions I learned to ask in the Huna tradition. Huna requires the practitioner to observe the philosophy of metaphysics; to take note of the outwards appearance of things and the nature of events as reflections of what is occurring within. Shinto also quite specifically asks a practitioner to observe that which arises in the natural world and reflect upon it as a lesson in schooling the self to discipline.

I recalled Bu's start in life and the words of his therapist that possessive and protective behaviours stem from his insecurity around valued items like his beds and favourite foods. Continuing to build up some trust and confidence in me was the key to a good relationship, trust in this sense, as referring to relying on me for a future action to help Bu.

Trust might also be thought of as having a confident dependence in the ability and strength of an another being. This kind of trust may create confusion when it is violated, or reliability comes into question, like someone else being on

Bu's bed, he did not expect. Trust though can be a means for finding freedom from the endless cycle of fear and desire. The Buddha has taught me that the only way not to be assailed by past and future was to be mindfully present, moment-to-moment, in your life, without attachment to the outcome of your or others' actions and reactions.

So, I asked him if I could share the bed, I apologised to him for invading his territory without permission. In less than a single heartbeat, he shuffled over to one side, without grumbling, giving me just enough room to make myself quite comfortable but be aware of his presence.

He had one more party trick left in him and it wasn't in this case, in relation to the spare bed. He took some joy in surprising people who came to the door, deliberately hiding around the corner, remaining perfectly silent until they came inside, where he would jump out from his hiding spot and bark excitedly in 'play bow' position, ready to engage. This surprised a few people, but they understood the game and played 'scaredy dog' by fake-barking back at him and making him run away. It was a kind of human stampede game.

A lovely bank manager lady had come to help me re-mortgage the house (yes for Bu's vet bills) and Bu popped out from hiding, clearly hoping for a human stampede game, but she ran down the hall as quickly as an Olympic sprinter and shut herself in the nearest available room. She was so terrified, she took some persuading to come out and trust me that he wasn't, in fact, going to tear her limb from limb and he was just having fun.

After a cup of tea and some restoring biscuits, tidbits of which I taught her to bribe Bu with, she got him purring 'BuBu, BuBu, BuBu' and all was forgiven on both sides. I did get the house re-mortgaged and paid off his vet bills. We headed down south about a month afterwards to visit Gary and entertained him with Bu tales of 'It's my bed'!

Bu's Lesson—Learn to trust so as to share the things that you love the most in the world – you will find someone prepared to curl up with you and be present with the snoring.

Chapter Eighteen
The Pack

I met Emma walking the bridle trail in the Perth Hills. Justin and I had Daphne with us, a dog we had fostered for a few months. She was a Whippet crossed with a Staffordshire Terrier, a nice balance of petite poise and waggly tailed engagement that gave her an enchanting look and hinted at a sweet disposition.

Emma was walking with her friend and her friend's dogs and we got to talking. Doggy people do this when we are out and about and we must admire each other's pooches, most of which are rescues. As a foster carer, I was always on the lookout for potential adopters. I did my best to entice Emma into taking her when I found out she currently had no dogs. It wasn't meant to be that day it seemed, although, I recognised the desire to have a dog in her eyes.

We had turned around to walk back home when I heard Emma running up to me, I thought perhaps she had changed her mind and did want to adopt Daphne!

Hooray! I thought.

Emma asked for my number to be in contact but not for anything to do with dogs it turned out. Her friend had turned her around and pointed her back in our direction and said, "*You need to talk with that woman, go and find her.*"

"*Why?*" asked Emma.

"*You need all the friends you can get,*" she'd said. It was the spark of what is still a close friendship, more than a decade later, centred primarily around everything dog.

I didn't re-home Daphne that day, but she was adopted shortly afterwards by a lovely couple who had lost their

Whippet cross earlier that year. They renamed her Bonnie for her sunny nature and the instant joy she had brought to them. It was rewarding to see all the dogs I fostered go to the loving embrace of their new owners and this was no exception. She was my second last foster dog and she had been in my care with my last foster dog Otto, a black and white Pit Bull and Pointer cross with a mischievous nature. They played hard all day together and after creating mild mayhem, would sleep together on a sling bed in front of the old combustion stove, whether it was on or not. It was hard to see them separated since they had become great friends but each one went to a wonderful home and I believe that their adventures ahead more than made up for it.

Bu arrived within a few months of finding homes for Daphne and Otto. Emma had too, in the intervening weeks, acquired a beautiful puppy, Tash, a Doberman and Kelpie cross with a gorgeous silky black coat and partly tan speckled 'wily coyote' nose. We quickly formed a pack, two dog mad humans and two human mad dogs. We took long rambles, enjoying sleepovers, snuggles *on* beds and teaching humans to give treats for tricks (not the other way around!) The early formation of the pack was made a little difficult by Bu being quite sensitive around a dog he didn't know but Emma was able to keep him calm by socialising them correctly. I recall sharing some stories with Emma about Bu's terrible behaviour and she passed these stories onto one of her teachers. Her teacher's response was that I should send him back to the refuge and get a better one that wouldn't cause such strife!

For the next couple of years, as Bu learned to be a confident and loving being, I wondered if the advice should have been taken after all. "*Who ever said it needed to be easy?*" I would think when dealing with yet another aspect of Bu's life. Practically anyone can take care of a good and well-adjusted dog, it takes tenacity and immense reserves of patience to look after a refuge savage. It's a drop in the ocean of rescues granted but let me here reflect on a statistic I read once from the Royal Society for Protection and Care of Animals. Fourteen million cats and dogs have been destroyed since World War II in Australia. I am glad that Bu was not added to this harrowing number because I did not give in or give up. In Zen there is nothing we are more responsible for than doing and saying nothing. Apathy seems so pervasive, so normalised that it's almost invisible to many of us. Non doing, not enacting compassion might be considered a subtle form of cruelty. Adopting a refuge dog, or any animal in need of a home is a small way of both saying and doing something and a bold attempt at pressing back against a tide of misery and suffering that plagues the planet. Love in motion arose to rescue Bu.

Campaigning to end the massive oversupply of healthy dogs and cats that end up euthanised is also a call to action I hope many more will hear. I hope it translates into loving action by ending the horror of puppy mills and greyhound racing, strengthening legislation for all breeders and above all, educating that compassion is required for each and every animal. Just because they are old, infirm and possibly damaged physically and mentally, does not make them useless and therefore disposable—it should instead make our hearts pound in response and give a healing home to them.

In my utopian world of a home for every beating heart, I would ban every single breeder, responsible or ethical or not, let all the refuges be emptied first and every dog and cat find a home before the next one is bred. I know it will never happen, but Bu gave me the courage to keep thinking that we can be saved through our actions, however small, and to keep up the fight against cruelty in any form.

For me being in a pack or community gave me the strength not to give in, not to give up on the fight and revel in the simple delights like watching dogs at play with their toys, tea in hand beside a friend equally as fond of the canine species.

Emma loved Bu too, we both had to be away from our beloved pooches from time to time and it was comforting to know they were in good hands in our absence. About three months before the end of Bu's life, I went to America. Emma lived at my house with the remaining pack and looked after everything. I had met a fellow vegan on the internet and wanted to explore the possibility of a meaningful relationship.

The meeting went well, he proposed at the airport—all very romantic but being still very much in love with Tarryn, I did not give it a really good chance and our engagement fell apart quickly after he had visited here some weeks later, flying away, hating Australia, my friends and me.

Coming back from the three weeks spent in the United States, I found my old beloved pooch quite faded in colour and in spirit and more than usually tired. I felt guilty to have pursued a romance which really was destined to come to nothing from the start in lieu of being present to my ageing dog. I realise now that I was searching already then for something to pre-fill the gaping hole, I knew his passing would leave.

Bu was excited to see me once he had woken from a deep slumber in his basket and I have the moment captured on video. I loved watching him come awake and seeing the light in eyes when he saw me. There is no need to wallow in regret of being parted for three weeks and limit my ability to experience joy. The exquisite joy of the moment of being together once more can only radiate in full if I am not suffocating it in fear and sadness. I think this is what is called 'letting go', the non-grasping that Buddha taught.

The process of ageing, a source of fear for many, can be considered a culminating adventure, up to which, the earlier years of our lives may be seen as a preparation for the opportunity to transcend the years lived with a defined sense of self. In doing so, we may transcend the fear of imminent decrepitude and death.

In this transcendence (here I mean being wholly embracing of the self with all flaws), we can dispense of self-preoccupation and ageing angst and be wholeheartedly available to serve others. We can still appear as a Bosatsu and Bu did this for me as I watched him age, going through the emotion of contemplating losing him with as little resistance as I could muster.

Some of the wisest words I have ever read are from a Tibetan Buddhist teacher Chögyam Trungpa:

"Let yourself be in the emotion, go through it, give in to it, experience it... Then, the most powerful energies become absolutely workable rather than taking you over, because there is nothing to take over if you are not putting up any resistance."

Emma went away for a couple of weeks too, I can't recall when, but we were all young and vigorous. I was looking forward to having Tash with us, I loved her slinky grace and was only annoyed when the squeaking got too much.

Tash was obsessed with squeaky toys in her young years, her favourite toys were rubber pigs that after much squeaking, tended to lose their voice when she ripped their throats out to see where the noises were coming from. At one stage, she managed to get the squeaker out in under two minutes—it was very impressive. Bu would then pounce on the abandoned remains and methodically rip them to shreds.

Tash could find a ball, no matter where we went, lakeside, parks, bushland, beaches—she *always* found a ball, as if she had some kind of ball radar on the end of her nose and once found, the throw and catch game would *have* to be played. She loved receiving toy balls too as well as squeaky pigs. A very well-mannered dog, she would always instantly accept these and show her approval by throwing them under her bed and playing hide and seek all by herself—knowing it was entertaining for us.

Bu loved a toy which we called the 'pimple ball', being a rubber toy with raised areas for cleaning the teeth and it had a tiny silver bell in it. The first one came from Emma as a present and he loved it so much, he slept with it in his basket many times but eventually, it broke. He excavated the bell out of the ball by ripping the perished pink rubber to bits and sat there bereft and crying once the destruction was final. I took us down to the pet store and thought how clever I was to find a blue one, a green one, a black one and a red one.

Bu's eyes lit up as I placed them down in a line in front of him, he paced up and down sniffing them, but he sat down and began to cry again!

I took off once more and went to another pet store, found a stock of *pink* pimple balls and bought all of them, knowing these things go in and out of fashion and might not be available again. I could tell he was immensely pleased when he found the seemingly resurrected ball in his basket from his happy breathing 'BuBu, BuBu, BuBu'. He fell asleep that night with it safely tucked underneath his muzzle.

I have the last untouched one still in a box with a few other favourite things; a blanket he liked to drag around and sleep on, a rope toy suffused with mint oil (it was supposed to help with his breath but it did not do the job with any degree of success), a brush with a few of his hairs attached to it and an adored soft Shar-Pei puppy toy, looking ragged with wear from Bu's slobbery attention.

We stayed a pack for around a decade and what a beautiful time it was. We are so different now. Tash is very old, Emma is spending every moment in care of her loved one, as I did with Bu. I have three young boisterous rescue dogs who are too much for Tash, who tolerates them as any grand matriarch should, with a mixture of reserved love and gentle warnings.

Bu's Lesson—Taking action with compassion reveals the preciousness of each life (and new toy) – in finding unconditional love we create harmony and happiness.

Chapter Nineteen
Sneaking Out

Bu had been very tired for a couple of weeks and seemed quite listless on his walks, as if he was only going out of daily habit and to please me. He would drag himself to the car, needing assistance to hop onto the back seat, where he would lie down until we arrived at our destination. I would help him out and he would set off at a slow pace with his head hanging down and within twenty minutes, he would stop and rest, stretch out on his belly with his head placed between his front paws as if to say it was all too much.

Bu had behaved like this before, many years in the past, before I moved to the Perth Hills. I thought he was getting ready to let go of life but the clever vet in Fremantle at the time did all sorts of tests and it turned out he had an underactive thyroid. Within a week of the right medication, he sprang back to life and was trotting haughtily around, terrorising the neighbourhood.

As he lay on the path once more, looking old and tired, I thought, "*Maybe his thyroid medication needs changing. A higher dosage perhaps?*" So, I dutifully booked a veterinary appointment, but a battery of tests showed he was perfectly healthy with no changes to his host of issues that we were managing. We were all most perplexed, but the answer came sooner than expected.

Tarryn and I were driving home from a lovely lunch out at a vegetarian restaurant, only to see a brave Nannup tiger like form tripping lightly up the hill, apparently heading for home. "*That looks just like Bomber…that is Bomber!*" Tarryn said, laughing. Bu's little ears were flapping up and down like little angel wings as he powered up the steep path, clearly pleased with his long adventure around the neighbourhood and oblivious to his discovery. No wonder he hadn't been so keen on his walks for a couple of weeks. He'd found a gate opening to freedom and took the opportunity to take himself off for sneaky outings.

I often wonder if Bu ever felt like he did not have freedom or enough of it. I don't think canines conceptualise freedom like humans do but they do seem to be more energised when on their walks, out of their home confines and if they are lucky enough, roaming in a nearby bushland or running along a beach. Watching Bu at play illustrated for me the importance of the Zen lesson of cultivating a clear mind as the absolute freedom, that is to say, not having freedom *from* anything.

I sensed Bu and probably all dogs experience living their life as these intelligent sentient creatures must do, but they perhaps do not over-think it and create mental angst around it. In this, they are liberated and have an in-built freedom.

Although, I do have to be truthful and say the look on Bu's sweet face was often one of great disappointment when he saw me close the lower garden gate, cutting off his escape route as the avenue to the freedom of his sneaky outings!

Bu's Lesson—Always choose freedom, that beckoning open gate—you never know what adventures it might bring.

Chapter Twenty
The Sky Is Falling!

There are beautiful and magical thunderstorms in the Perth Hills, the sound is incredibly loud and shakes the house, rattling the windowpanes. Lightning is something worth getting out of the warm cocoon of a bed for, stand outside and look at it striking in the distance, lighting up the stunning view of surrounding bushland.

Bu did not have such an appreciation of thunderstorms. He was so nervous of them that if I knew one was on its way, I would gather up my bag and leave with a laptop to work from home or cut short a visit to a friend. He needed someone to be with him as he explored places to be safe—under the bed, under the covers of the spare bed, outside under the house, an ash filled fireplace—nowhere was that comforting for him.

When I was not home, Bu would head out of his dog door and stand at the decking gate, whining for a bit, and chew the gate uprights as if deciding what to do. He rarely jumped but when a big thunderstorm passed through, he managed to clear the gate and would head immediately to the safest place possible—the post office, about two kilometres away, down a steep hill, across a busy road, where he would, so I was informed, bash on the door to be let in.

There, he would sit under the counter, shivering and shaking whilst I got a call to come and collect my dog. I was amazed when I went to get him that he had achieved a remarkable truce with the resident cat, a huge black and white moggy that oozed all over the bench, taking up at least one station, such was his size, he looked as if he owned the place.

He tolerated the wet and afraid interloper with that peculiar disdain that felines seem to have for anything other than their own species, and looking down with his superior gaze on Bu as he walked out with me, seemed to say 'Good riddance!'.

I collected Bu from the post office three times and as for the chewed gate, the marks are still there, and I won't be removing this reminder of his torrid anxiety in response to the thunderstorms and it dissolving him, like being at the vet, into a 'wee timorous beastie'. The dents in the wood are too precious to obliterate, when I catch sight of them, I smile for the beautiful dog who thought the sky was falling and how there was always help at hand for him in my absence.

My brother loved Bu and to distract him once, started patting and playing with him by pushing all the wrinkles on his head forward, towards the tip of his nose, saying in an hilarious fake Spanish accent, reminiscent of Manuel's in 'Fawlty Towers', 'outside brain dog'. He would then stretch the wrinkles all the way back to his neck with 'now he's hamster'. Bu truly did look—in this playful interlude—like a dog with a furry wrinkled brain on the top of his head and then, exactly like a worried hamster.

Bu would end up wriggling with delight, flop on his back for a tummy rub and forget about the sky threatening to fall on his head outside. I loved watching them interact. Bu had a very soft spot for my brother and actually knocked people out of the way to get to see him as he came through the door on a visit.

I suspect to this day, that like my friend Gary, my brother would smear meat paste behind his ears, ready for these visits. Bu would go straight for them as he bent down in greeting and give them tiny quick licks until it tickled too much. My brother would entice him too with other interactions. Making his tail go round and round in fast circles. "*Schmoopy Schmoop Schmoo*" he would say, who is a "*Schmoopy Schmoop Schmoo*"?

The expression 'Schmoopy' came from an episode of Seinfeld where he and his girlfriend at the time are being sickly sweet to each other, using this as their mutual endearment term. It suited Bu and my brother somehow, one big brave dog and one big courageous man, romping around in unfettered fun.

Bu also was very fond of my father to whom he always gave a special greeting at family visits. We based our visits on going around a lake and its wetland surrounds, situated quite close to my father's house. It was a daily occurrence for a time when we all lived close by. The dogs would pile into the car and we would go out to wash off the day's distractions.

Watching Bu snuffle at the water's edge, I was surprised to see him foray in a little further, peering at something a few metres away. Then, risking wet paws, he proceeded into shallow waters and waded right in to investigate what turned out to be, when it popped its reptile head up, a turtle, that having been disturbed was paddling away from him. Suddenly, Bu was trying to swim after it and keep his head above water, but he was not designed to float. He sunk quite quickly and whilst doing so, spun around and ended up buried in the algae laden mud of the lakeshores with only part of his rump and the tip of his wagging tail showing.

He executed a smart commando crawl out of there, a picture of rapture in slime. All we could see were two little eyes blinking in a face of silty gumbo. We laughed so hard, we bent double, the sight was so unexpected and funny—like a scene out of a cartoon.

I recall it as if it just happened and I recall a Buddhist monk once taught me about different expressions and forms of love. In essence, he said that the truest expression of love is when we pay attention to someone or something and we are not distracted, invaded and carried away by our own thoughts.

Bu gave me an understanding similar to this in what I experience as a *suspended moment,* a moment of deliberate pause in which to inhale the potency of it, like witnessing the raw power of lightning in the sky and feeling the grace that is instilled in it, like holding a trembling Bu against the terror of it falling. In watching and laughing at his swamp antics and taking him home to wash him clean—under protest of course—Bu thought he smelt heavenly.

These ordinary moments of our lives are distilled and captured in my heart, a piece of time remaining incorruptible through the most powerful force in the universe, *love*.

Bu's Lesson—When the sky is falling on you, there are many sources of comfort and help – let them play with your wrinkles and demonstrate pure love.

Sniffing Bu

Art credit to: Nada Orlic

Chapter Twenty-One
Encounters of the Strange Kind

When my husband Justin and I lived in the Perth Hills on the property with all the rats in the palm trees, frogs in the winter creek and possums in the roof space, there were plenty of delightful walks to take Bu on. We could simply step out the door with a prancing Bu (he loved going for walks) and head in any one of many directions. The walks ranged from three-kilometre strolls to about a six-kilometre mini hike and our path was dependent only on how much time we had and how much energy Bu had.

All the walks were studded with their own unique native bushland features of granite outcrops, stunning stands of old jarrah, wandoo and marri trees. Underneath were bright green and red kangaroo paws, ancient xanthorrhoea trees and in spring, a riot of splendid coloured flowers including rare species of orchids hidden beneath the bushes, unless you looked really closely.

It was a gorgeous day when we stepped out with Bu for a small stroll, we wandered down the road to go into a secluded part of the forest, and although close by, decided to keep Bu on the lead just in case he decided to frolic unwanted in someone's yard. Like most dogs, he did enjoy the thrill of the forbidden and would take time to sniff *everything*. He had 'the wind in his tail', as my Mother used to say, as we set out—clearly excited to be in the fresh air, even though he had left his comfortable place in front of the warming combustion stove. Exactly where the two foster dogs Daphne and Otto used to curl up in their sling beds.

We walked along talking and passed all the quirky houses, and for the first time, saw in a paddock, a herd of llamas. The llamas were peacefully grazing and chewing cud, only a few turning their heads looked mildly interested as we went by. Then, a large male llama spotted Bu and raced to the fence to investigate. He made a ferocious humming sound as the rest of his flock gathered around him to stare down at the little barking creature, they also started humming and stamping their hooves on the ground. Bu stopped barking as he became entranced by the intensity of the two sounds, a high-pitched humming counterpointed to the thud of the hooves, and he simply sat down with his head cocked to one side to listen closely.

The chief llama had decided whilst Bu was not a threat, Justin was, and with unerring accuracy, spat the contents of his belly directly into his face. Justin was covered in foul smelling and mucky looking green goo. It was the second time I had seen this. Many years prior, my first husband, a beautiful American man called Todd, had tried to pat a llama at a petting zoo down in the South West of Western Australia, close to where Gary lived. Todd was very handsome, very muscular and spoke in a sexy southern drawl—even when provoked—and was, and is, incredibly kind to animals. We had a great relationship but years apart with his work and my work created a distance too vast to traverse our way back. The sweetness though, lies in that we are great friends still and always will be.

He too ended up with a face and shirt covered with llama spit only moments after I had been petting it and cooing sweet nonsense into its beautiful pointed ears. I recall too on the same day, poor Todd had gotten low to the ground and fed a green apple to a miniature horse who nuzzled him gently after consuming it but as Todd turned away, took a huge bite of his retreating back.

I had been teasing Todd about the venomous snakes, poisonous spiders, nasty eye pecking out magpies (he had been swooped that same day), to watch out for drop bears, and he told me, in his relaxed southern style, he couldn't wait to tease me in return when I got to America.

Which he did, without any mercy, successfully scaring me with wild tales of coyotes, rattle and sidewinder snakes as we walked in the Morongo Basin near our home. All of these glorious critters I saw at one time or another and recalling his stories, would startle as they emerged from the desert dunes. The scary animal encounters faded into insignificance when roadrunners would burst out of the sparse scrub, running fast in their strange bent up 'V' form, the kangaroo rats peeped out of hiding holes at dusk and the magnificent sight of a giant desert tortoise ambling along.

I treasure the times we spent together, Bu not yet part of my life, encountering weird and wonderful wildlife. As I said, there is still a deep abiding friendship between Todd and me and this, as well as the memories, are to be treasured. Todd is one of the most peaceful and considerate people I know—I am sure there is a Zen master hidden inside him—and he is incredibly kind to animals as you'll discover in the story of Gromit. (See Book Two, Chapter 1)

Strange encounters always happened on the walks that I took with Bu around the property where I live today. One day, Bu and I took an autumn stroll down the hill towards a creek line where we would cross an old-fashioned wooden bridge and take ourselves onto what is known as the 'bridle trail'.

We both smelt something in the air, Bu was dancing a jig on the end of his lead and I was looking around for the source of the pungent smell when a large black pig, with a reddish collar on it, jumped out from underneath an apple tree, having feasted on the windfall and trotted down the road.

It surprised both Bu and me, we both stood there amazed, our usual experiences on that road had been kangaroos hopping down the road, possums scrambling up into trees with handfuls of stolen fruit and rabbits bouncing around. We followed the pig and heard its owner calling to it as if it were a dog, and it responded just like a dog. Its head rose up quickly, its little curly tail stiffened, ready to run and off it went to its house for a loving scratch behind the ears and under the chin and what looked like a healthy breakfast. Bu looked fascinated but as he had a history of nibbling on other dogs and other creatures, we moved past and in search of more mundane encounters.

For the first time ever, I had some consulting work requiring travel to the West Australian wheatbelt. I took Bu, who loved a car ride and had been invited along with me to stay out on a wheat farm. We passed huge properties and saw all kinds of animals, cows, sheep, deer and emus safely penned up in dry looking paddocks. I found the sight sad and disheartening, although they were all well cared for. Bu was in bliss with his little ears and lips flapping in the wind as he stuck his head out of the window.

I pulled up into town to get some breakfast supplies and get some information about the place. It helps to orientate me on these assignments if I know some history and can get a glimpse into the culture. It was a quaint town, old buildings with deco facades, museums run by local volunteers and friendly people who said hello to strangers. What I did not expect, walking through the town I was to work in, was a lovely meeting with a goat called Nanny.

Nanny was a rescue goat, raised by hand, having been orphaned at birth. Nanny was on a lead and whilst wary of Bu, her yellow slanted eyes were alert for impending mischief and she kept sauntering along, occasionally nudging him as he sniffed her, both quite undecided about what to do. I was told she liked to go running with the dogs in town. We kept walking but sought out a cafe for Bu to sit down at. When we found just moments later, a gorgeous little bakery with excellent smelling coffee, he parked himself at the nearest table and looked up at me as if to say, *"Let's not keep going, it's way too embarrassing to be seen with a goat, so let's do lunch here and just forget about what just happened."*

We did just that, peacefully observing Nanny's progress down the street as I had a delicious lunch and Bu chewed on a bone, kindly given to him by the cafe owner who much admired his shiny brindle coat saying, *"he could pass for a Nannup tiger you know…"*

Bu once had an up close and personal encounter with a feathered Australian native. We had emerged from a bush walk near Gary's place and slightly disorientated, had ended up near a blue gum plantation bordered by a pine plantation and a river. At this lush intersection, a whole family of emus were peering at us short sightedly, no more than three metres away and separated from us by a short wire fence.

Bu raced up to the fence and jumped up, resting his paws up on the top rail. The emu looked at him, turning his head from side to side as if weighing him up. With three short steps, it started pecking methodically, repeatedly and without any sign of fear, at Bu's forehead. Bu simply blinked with each sharp peck; he didn't budge. The emu dad, surrounded by five of the most adorable emu chicks, about the size of a watermelon, kept right on pecking and still, Bu would not budge—he seemed hypnotised.

They soon came to an impasse; the pecking wasn't doing anything, and Bu was not about to back down in a hurry either. Emus are quite docile unless provoked and are known to be curious, just as Bu was. Curiosity satisfied, they both turned their backs on the other, Bu headed for home to Gary's with me and I watched entranced as the emu went down to the river and played in the quiet pools of water, flipping himself on his back, waving legs and long toes in the air as his chicks waited on the muddy shore.

What a sight it was. If it could have, I think Bu's jaw would have dropped. Mine definitely did. It's not a sight I have seen since, although research tells me it's common for emus to do this when the weather is warm.

Possums have inhabited the roofs of both my Perth Hills properties, and it drove Bu to distraction when they arrived in the winter months and danced heavy footed like they were having a possum disco on the roof from dusk until dawn. They then spent the spring months hanging out in the nearby trees, throwing things down at him and chattering away, taunting him. He would watch helpless to do anything but bark and slink away to his sandpit to rest his weary doggy bones.

They were lovely brush tail possums in both houses. It was charming to see one of them (I can't recall in which house it was) climbing the gum tree, with a baby possum on its back, for a few weeks. We had seen a similar sight at Gary's, but that was a pretty ringtail possum, a very confident ringtail possum in fact. When we had gone to bed, the possum would come to the glass sliding door triggering the sensor light. The cheeky creature's form illuminated, it would pace up and down the outside until Bu would slide off his couch with reluctance, appear at the screen and go nuts! Barking his head off, he could not get to that naughty furry thing outside, parading up and down, staring in at him with its huge eyes, teasing with a few flicks of a ringed tail. Oh, how he hated that possum!

Bu's encounters and events were both very funny and profound to me as I recall the many and varied colourful characters (other than Bu) who have left their indelible prints on my life experiences. Funny in that they make us laugh and profound in that these things seemed to occur just by chance and enriched our relationship.

As referred to earlier on in the book, the eminent psychiatrist Carl Jung coined a word for these kinds of meaningful events, chances or coincidences—*synchronicity*. Jung postured in 'Synchronicity' that events could be meaningfully connected, not just by occurring simultaneously in a normal cause-and-effect way, but in a way that reveals something profound about the deeper and unseen structures of the unconscious mind.

Zen followers observe meaningful events as synchronised too and further, as a very clear mirror that reflects what you already believe at an unconscious level. If you believe in kindness and compassion as force to move the world, you will see many examples arise, one after another, to prove your belief right. The more events like this that occur, the more it is seen as a mind open to moving towards Satori or enlightenment, where we might experience a sense of oneness.

During quiet moments, like in meditation, we may lose the sense of separation between self and other. This I consider to be a synchronous event, experiencing it as a tiny glimpse of Satori. When you see your reflected thoughts and emotions in the other, this is direct experience that you are not limited to merely the individual self that you normally think of.

You are seeing a small demonstration of the fact that there is no real separation between you and the world around you. Bu and I were connected with everything in the bushland and all the encounters were reflections of a conscious craving to be bonded with other.

Bu's final meeting with wildlife gave me such joy as I had been wondering when I would ever see one—I'd been bushwalking for three decades without seeing that elusive hedgehog-like native.

The very first encounter I had with an echidna was with Tarryn and Bu walking along a firebreak around Nannup. The Nannup tiger was in fine form, snuffling through all the undergrowth, making his way through the overgrown scrub with slow but thorough progress.

He stopped at a fallen marri tree that was exposed to the firebreak at one end, down his head ducked and up his butt went, the movement punctuated with little snorts of excitement and high-pitched yips. He kept circling around the end of the tree, trying to get to whomever was hidden beneath.

We leashed Bu quickly to keep whatever it was from harm, and out a short-nosed echidna crawled, spines bristling, still licking a few termites (they need to eat a staggering forty-thousand ants and termites a day to survive) from around its mouth. We had disturbed it at mealtime. It waddled slowly across the red pea gravel in front of us in its strange splayed gait—seeking out a quieter burrow and no doubt, an uninterrupted lunch.

The spines are sharp on echidnas, providing a natural deterrent to any predator but Bu was tugging strongly at the lead and looked like he would not mind risking a few in his snout for the intrigue of a new encounter of the strange kind.

Bu's Lesson—Seek and you shall find the weird and the wonderful and the dear companions along the way to help you celebrate it all – it won't be by coincidence either!

Chapter Twenty-Two
Scavenged Prizes

Bu loved a scavenged prize, the more decayed and smelly, the better. He was a gentleman canine however and gave me a continuous stream of rotting carcasses as tokens of dog's undying affection; he always presented his finds to me at my feet, wherever we were.

Over a decade, a supply of deceased, in increasing order of magnitude, of no less than three bush rats, more than a few stumpy tailed lizards, a couple of magpies, one dugite snake, a rabbit or two and much more than I can clearly remember, dotted our relationship. Two presents do stand out more than others.

With a large grin on his face, Bu ran full pelt out of bushland with something clamped between his jaws. He weaved and ducked with what appeared to be a very long and heavy stick in his mouth, which as he drew closer, turned out to be a long since departed emu leg. He wasn't quite aware of the full extent of the length of it though and attempted to run between two young jarrah trees only to be caught between them. He was stopped rather suddenly and vaulted over the top of his prize, landing on his chin, paws spread out in front of him with a heavy 'ooff' (again). It's lucky Bu was built like a tiny tank, he shook these incidents off rather quickly, picking himself off the ground with as much dignity as he could muster with the wind knocked out of him.

He ran around, fetched his emu leg and went the last few metres towards me to drop it at my feet. I praised him for the find which was so old as to be entirely dry and leather like. I gave it back to him, as was our tradition and he lay down to gnaw on it a bit. I used these moments to find a rock or log to sit on and contemplate the beauty of nature, my fortune in knowing Bu and the infinite mysteries of love.

He did not stay long, the emu jerky was not to his taste, not smelly enough to be bothered with I guessed, so, we went home where he excavated from a potted plant, a chicken carcass, stashed from the day before, as a consolation prize.

A grislier scavenged prize was presented to me when exploring some thick bushland around the picturesque Margaret River wine region of Western Australia. Tarryn and I had parted ways, I was feeling raw and sensitive staying by myself with Bu at a dog friendly cabin.

I felt my mind, body and spirit crying in unison of longing for things to be other than they were. The grasping at the ephemeral threads of memory was a moment-by-moment exquisite torture. This kind of state of disconnection with reality and internal chaos causing suffering was far from peaceful. I knew this, I experienced the separation as a kind of melancholic madness. As with anger, I did not try to change the emotions. I simply experienced them, allowed them, and then tried to find a place to settle and turn down the dial on the suffering.

A tendency we can have as humans is to lock down and wrap a cocoon around our pain—shutting the emotional door to the world, and barring it against a grief that feels overwhelming, too much to bear and too ugly to show. It's that lock-down that gets us stuck, so we get convinced that we will never let go, maybe never be happy again. Not a place to settle into if we want to move through suffering and beyond to a place of ease.

So, I went exploring the lovely new surrounds of the cabin, the beauty of it helped me cry out the last tears of pain, locked away inside. The tears had cascaded fast, each one landing with an audible 'plop' on the path only to be obliterated by the fall of my foot. Their wet imprint only momentary at best.

With each step, my tears ceasing, I felt a return to being suffused with peace, a feeling that needs to be present to begin accepting loss, and being apart from Tarryn had felt like a huge loss at the time.

The walk took us about an hour and a half with me shattering the pain with meditations on impermanence as I walked. Bu bouncing around me, would look up concerned but seeming to say, "*I know you really, really loved him… I'll miss our boy time…but look, there are bushes to pee on, small trees to bulldoze and dead things to find for you…so get a grip mum, it's your thinking that is causing this grief.*" He was an amazing communicator and just a single look conveyed it all.

From many wisdom traditions, not just Zen, I have gleaned that strong emotions don't have to cause us to feel like we are helpless in a riptide to get drowned in; they can wash over us like waves. American Tibetan Buddhist nun Pema Chödrön in '*When Things Fall Apart*' says:

"*The only reason we don't open our hearts and minds to other people is that they trigger confusion in us that we don't feel brave enough or sane enough to deal with. To the degree that we look clearly and compassionately at ourselves, we feel confident and fearless about looking into someone else's eyes.*"

To keep an open heart in the time of separating from a beloved partner or losing a loved one like Bu through death is the greatest and most rewarding challenge of all. It comes with the alluring prospect of the prize of 'Satori' or enlightenment—a prize of a much different sort to the one I once received.

Bu had been missing for a few minutes and calling to him to come, I realised that just on the edge of my hearing, I could hear something heavy being dragged through the thick undergrowth. The sound got louder and I turned to see a striped furry bottom coming towards me, more or less in the air, moving backwards towards me as it yanked something closer. It looked like hard work and it was as bad as I had imagined when he reached me.

Bu was dragging a *whole* kangaroo carcass, clearly a victim of a road accident and not that fresh from what I could tell, with the maggots that were crawling all over blood-soaked areas, the spilt intestines trailing out from its belly and the stench wafting from it. It had been in life a rather large and beautifully marked male, a white star on its forehead and a blaze of light grey amid the usual dull brown on its chest and only Bu's solid frame and sheer determination had allowed him to bring it to me.

Of all Bu's finds, this one was the biggest and the most pungent. Thank you and well-done Bu! After quickly praising him, I broke with our tradition of Bu sitting down to feast and meditate. With a lot of encouragement, we were able to walk away, with Bu soon far ahead on the track in search of more prizes to scavenge and show his adoration of me.

Bu's Lesson—Everything is impermanent; life, love, scavenged bush dinners – the thing to know is how to savour them deeply while you have them and to let go into the unknown next.

Chapter Twenty-Three
By Any Other Name

Any dog in my family has always had quite a few names and they all had reasons to come into being, often from funny events, but I think Bu had more than his fair share of them.

Bu variously was named by the people he loved through his funny moments, like stalking through the backyard grass and hunting for pesky rats around the shed. My brother-in-law at the time noticed he was picking up his feet and bending his back knees at a weird angle (to avoid the dew on the grass, lest they get wet) and began singing '*Knees up Bomber Brown... Knees up Bomber Brown*,' making a quaint parody of the song '*Mother Brown*'...

> *"Knees up Mother Brown*
> *Knees up Mother Brown*
> *Under the table you must go*
> *Ee-aye, Ee-aye, Ee-aye-oh*
> *If I catch you bending*
> *I'll saw your legs right off*
> *Knees up, knees up*
> *Don't get the breeze up*
> *Knees up Mother Brown"*

'Knees Up' being an expression I learnt from my English mother as a word used for a dance or a party and it fitted Bu so well, being always ready for action and the life of any party. The name stuck and many people outside the family who heard it and who loved Bu, sang it to him to entice him up out of a comfortable bed and get him excited to play. He was always happy to be enticed like this, grabbing hold of someone's hand affectionately and playing the 'bitey bitey' game, daring them to wrestle him harder.

When Bu sat up, looking out a window at the world outside or in contemplation of deep doggy matters amongst the bushes, he would look for all the world like a bear. Perhaps it was his wrinkles that collapsed like a saggy concertina and gathered around his neck and maybe it was his tiny ears, but it led to him being called the Honey Bear or Baby Bear.

His magnificent tiger stripes led him to being called the 'baby tiger', the 'Nannup tiger' (if down in the South West of Western Australia, to those who knew him) or the 'brindle baby'.

My brother had, as you know, a way of distracting Bu from being in terror with "*Who's my Schmoopy Schmoop Schmoo.*" It made Bu wriggle in delight and of course, he would start to snort and puff 'BuBu, BuBu, BuBu' with the joy of being touched and loved—so, he also was known as 'Schmoopy Schmoo'.

Bu used to frequent the compost bin of whichever house I was living in. This was a must-have and the first installed thing in the garden, which he would know somehow would be ready for him to check in around about three weeks. He trotted up to the huge bin with some ceremony on 'Compost Day'. Inspecting the outside thoroughly, he would place snout to plastic at the base and then hop up on the lid itself in the hope, however vain, that there was tasty, mucky overflow to snuffle up. Snuffle is a real word, truly...being the correct name to use for the long furry trunk character Mr Aloysius Snuffleupagus from the TV show 'Sesame Street'.

If anything had overflowed, it was from one of the air holes, generally it was a pumpkin vine that had emerged through and was flourishing in the nutrient rich environment from a discarded seed. I lost the picture I took amongst one of many computer disasters, but I can see clearly in my mind, a small brindle dog, standing amongst a forest of head high pumpkin vine leaves, cropping them like a small cow, leaving rounded bite marks in all of them. This is how Bu became known as the 'lesser known brindle cow'. I am glad it was him, as I had at first thought I had a massive snail problem when I saw the munched-on leaves!

For surviving his many misadventures, illnesses and injuries, he was known as 'Wonder Dog', after the cartoon character in the Super Friends television series who escaped many serious situations and was very loved as a sidekick to the two superheroes, Wendy and Marvin.

For his immense strength to body weight ratio and sheer stubbornness of being, Tarryn named him after a bulldozer brand 'The Komatsu'. To me, he exhibited a small tank like tenaciousness, rolling through the undergrowth without pause, in pursuit of his quarry, so he was known too as 'The Tank'.

It's not polite to tell you the other names he got called as he stole cheese off brunch boards, took chunks out of steaks being transferred to dinner plates, was discovered sleeping on freshly laundered sheets, when he dug up potted plants in back yards (not his own) and tried to take a nibble out of someone he didn't like the look of who was trying to sell something to me at my door (fair enough, as I didn't need a year's subscription to whatever it was). Once he was caught red pawed with a whole roast chicken clamped between his jaws, on the new cream coloured carpet, at a friend's house. He had stolen it off the kitchen bench while it was resting, the temptation too much for a dog of skilful means. It would have required the patience of Buddha himself to restrain myself from letting those unspoken names escape my furious lips.

I initially began my stop and start study of wisdom traditions like Zen, to study myself, to know who I am and the true nature of mind from a non-religious and non-scientific view. I like how Zen addresses practical matters, such as how to lead a good life and how to handle difficult challenges by understanding the mind. What I find now is that the berserk, chattering and, dare I say it after years of examination, the insane mind, can find utter peace through meditation and cultivating Bodaishin, the practice of loving kindness. It is through these practices that we can get to know exactly who we are.

Bu reminded me that in each of his names, his true essence was contained and was always integrated in every aspect of himself. The whole of Bu was for me a marvellous Bosatsu. Turning up each day, simply responding to all names when called, he was a dog who, without struggle, knew *exactly* who he was at *all* times.

Bu's Lesson—Go ahead, and enjoy naming all the aspects of who you are – someone, maybe even you, will love each part if not the whole.

Chapter Twenty-Four
It Wasn't Me!

One of my favourite things (Bu had many and he'll tell you soon what they all were) was to sleep at the farm in Tarryn's caravan. Set amongst the great marri trees, on waking, I would look out on the paddocks of cows and hear the cries of the ibis and the red-tailed black cockatoos as they wheeled overhead to come to settle on the skeletons of the dead trees near the dam, or alight on the troughs to drink. In the morning, we would hear the chickens scratching around behind the caravan gently clucking to each other from the other side of the fence. It was idyllic, rural peace.

Generally, Bu would still be asleep in his farm basket (another bed) set on the floor and continue to serenade us with doggy snores until the light penetrated through the windows and woke him. If it was a hot summer, Bu would happily sleep outside and enjoy the night breezes. I would get up and check on him as he got older to make sure he wasn't getting too stiff with cold. This gave me the chance to look up at the stars, which that far out from Perth and with no light pollution, were stunning in their clarity and closeness. I often saw shooting stars and foolishly would make a wish to capture the moment of peace and love forever.

I had checked Bu early in the morning and gone back to bed for a little more sleep. About an hour later, Tarryn and I were woken up to the sound of a chicken in distress. We peeked out the windows to see Bu running around, chasing a chook that must have flown over the fence into the caravan compound and was now hop-flying to safety. As I ran out, I saw an uncharacteristically athletic Bu jump up and with a snap of his teeth, grab the chicken out of the air and sit down, most surprised he had actually caught it!

He glanced up at me and spat it out of his mouth with a sideways flick of his head as if to say, "*Chicken? What chicken?*" He proceeded to also spit the feathers clinging to his mouth and face away with puffing noises—"*These feathers clinging around my mouth…no idea!*" By this time, Tarryn had come out to witness the incident and we collapsed with laughter as he sat there, trying the innocent—"*It wasn't me!*" look—for the crime just committed.

The chicken was fine, minus a few tail feathers, having flown back over the fence and immediately surrounded by the gentle clucking of her concerned companions. We kept checking on her throughout the day, bribing her with kitchen scraps to come close. She was back to normal very quickly after her terrifying encounter with Bu.

The farm, you may have gathered, is an incredibly peaceful place and I felt restful whenever we visited, no matter how short or long we stayed. I felt so well and content wrapped in the fresh air and in Tarryn's arms, knowing that my furry companion was enjoying the chance to play with the cows and play doggy games with the resident farm dog Jemmy. The only illumination at night was the moon and stars. This was intoxicating and I felt so fortunate, and do to this day, that I have these moments to cherish. The stars are spectacular where I live now but I no longer have Bu to check on in his slumber so it's only a lightning storm that would rouse me out of bed to admire their brilliance until dawn. I guess at the farm, there was nowhere to go, nothing much to do in particular, and the opportunities for stillness were ever present. Meditation came effortlessly in this idyllic setting and it was a period of cultivating stillness and mindfulness that has carried me through the intervening years.

Zen meditation is the type of stillness practice I am most familiar with, although there are myriad techniques that work to a greater or lesser degree for me. I return to it like an old friend and find that my relationship with this practice, as I grow older, has deepened.

The resistance to sitting still has lessened for the most part, although I notice that when I have been writing about Bu or thinking about Bu, it's more difficult to calm the agitated state of my mind. Ever the Zen dog, he is still teaching me!

This is the essence of the practice, to show up and do it, no matter how tired you are, no matter how sore you feel and no matter how in flow or out of flow you feel, meditation is the prescription that will elevate the mind to a state of peace and even bliss. The chaos will fall away gradually as you sit—at least that's my experience and I would highly recommend that even if you have a great teacher like your dog or cat, the benefits of meditation are so many, it's worth finding time each day to sit and do it.

With the feeling of calm, there also comes the feeling that a great well of loving kindness can be drawn upon and offered up to family, friends, anyone in need really that crosses your path and of course, creatures like Bu, who will give back unselfishly for any small thing that they have received from you.

Bu's walking meditation for that day was to try to achieve a second chance at that feathered escapee by patrolling the fence line, ready to pounce. No low flying chickens appeared, so eventually, Bu took up his favourite spot on the caravan deck, head resting between his paws and closing his eyes. He appeared like a Zen master, meditating in the soft sunlight, breathing slowly in and out, 'BuBu, BuBu, BuBu', a rhythmic puffing of serenity a counterpoint to the magpies singing their delicate melodies in the jarrah trees far above.

Bu's Lesson—For an opportunity that may have seemed out of reach, just meditate (or mooch) on it – you may just be surprised at the outcome of catching it.

Chapter Twenty-Five
Tin Legs

I mentioned Jemmy earlier, the dog who lived on my friend Bruce's farm in Gidgegannup. Bu adored Jemmy and they were quite close. This was evident when I drove down the long driveway, passing flourishing olive groves, sheep grazing in the paddocks and huge peaceful looking Buddha statues set under flowering gum trees. Jemmy would race up from whatever she was doing, barking in greeting, and dancing in front of the car for me to slow down. I would have to stop so she could jump up on the door and say hello to me. I would then open the door and she would immediately stick her head into the back seat and offer sweet little doggy kisses to Bu who would always return the love, giving lick-kisses right back, his stumpy tail going '*thump, thump, thump*' on the seat. His eyes somehow grew softer whenever they met, I swear they changed colour to a light-burnt caramel from dark brown, as if seeing her somehow illuminated him from deep inside.

Every dog lover learns to read their own pet very well and although he was rarely vocal, his eyes would tell me everything I needed to know. He would communicate silently when he was uncomfortable or in pain, his eyes would go from their normal round lustrous liquid pools of love to tight wedge shapes with all the shine gone. Wrinkles would also form on his head if he felt profoundly uncomfortable. I knew when it was time to offer medication and massage, a magic combination for Bu, who would soon be up and about, doing important dog stuff.

He did important dog stuff with Jemmy often, rounding up the free-range chickens, finding and eating their abandoned eggs, hunting uselessly for rabbits (farm rabbits are very lean and fast), splashing in the dam if it was hot and seeking warmth in the winter inside the caravan—a rare invitation for Jemmy.

One lovely winter morning, having stayed the night in the heavenly seclusion of the caravan in Tarryn's company, I was seated on the grass with Bu at my side, both of us eating the same healthy breakfast of porridge, fruits and seeds. I watched as Jemmy came up to the fence for her daily visit and was surprised to see Bu stiffen up and rush to the fence, hyper alert, in a posture I had never seen him adopt before—he was thirteen at the time. I thought I had seen every mood. This was something entirely new and I thought momentarily that maybe he was going to attack.

Far from it! Jemmy was all but swooning as I opened the gate, she raced in, frisked around him like a puppy and proceeded to rub herself up and down his flank, turning around to present her furry rear end to him and then I understood—she was on heat!

Bu wasted no time at all in making it clear he was *very* interested to give it a try, despite not having, shall we say, the wherewithal, and certainly no experience in matters of pup production. I don't think I had ever seen Bu quite so intent on anything and quite so oblivious to the world as he tried, no less than *seventeen times,* to make up for lost time. The poor dog was exhausted halfway through this count, but when he was separated from her, they both cried so hard, pressing up on each other through the wire fence we laughed and let the romance continue. Bu ended up so tired from his efforts that by the end, he was simply hanging over Jemmy with his tongue protruding, tired but clearly a very happy old dog. Old dogs can learn new tricks!

Jemmy was a little bit smaller and having had her fun, decided to shake her muscly beau off and seek out other farm attractions.

Jemmy's owner, Bruce's son, arrived just as this hilarious interlude in farm life was over but not having witnessed it, came up to Jemmy to give her a huge pat, which he did and looking over at Tarryn and me, said, "*Eeew, she's all sticky.*" We informed him of the source of this, needless to say, he was horrified and leapt about, with his hands waving madly in the air, saying, "*Yuk, too gross.*" He went off to clean his hands, Jemmy trotting by his side, looking the picture of girly innocence.

Jemmy did not appear again until the next day when I was packing to go home. Bu got up very slowly, hobbling over at a snail's pace to say goodbye, looking for all the world like his back legs were made of tin.

Bu knew how to suck the marrow out of the bone of life. As I lifted him into the car, too stiff to jump up but replete with farm experiences, and me, full of love for the moments on the farm with Tarryn and Bu, I recalled a poem a great friend, who also loved Bu, had taught me, by Dylan Thomas, '*Do Not Go Gentle Into That Good Night*':

"Do not go gentle into that good night,
Old age should burn and rave at close of day;
Rage, rage against the dying of the light.

And you, my father, there on the sad height,
Curse, bless, me now with your fierce tears, I pray.
Do not go gentle into that good night."
Rage, rage against the dying of the light.

At thirteen (about seventy-four in human years), with this new encounter and sheer focus on the moment, I felt he was setting us all an example to live life to the full. A life, resonating in harmony with the natural order of things and not in constant conflict or unease with it. This, perhaps, is one of the important essences of Zen, being fully present to each fleetingly moment, whether we are washing dishes, hugging a family member or working in our chosen occupation. If disenfranchised from the moment by wishing that it was otherwise, that we are elsewhere, with someone else, then we risk wasting the potential in each precious individual life, perhaps disconnecting from not only our true nature, but others around us.

My many teachers, from various traditions have taught me to be 'at cause' of my life and not at effect. This simple, yet profound lesson has led me to try to be the best self I can be and operate, not only in a mode of loving kindness, but as if life is a great adventure. We can only expect one outcome as a human that is certain; that we will pass away. In-between it is up to us to be present to each rich experience we draw with intent into our life.

'Don't live just to die' these teachers also taught me. To live life to the fullest, discard fear, embrace the new, which when we do, is exhilarating. We feel truly alive, tingling with anticipation that brings focus and clarity. Reality from this mindful state then is negotiable and it depends on an individual's perception. To feel the tingle, to seize the moment, is to live in a reality of your own design. Designed well, it can be a source of inspiration, peace and love, a true adventure.

Bu was the master of Zen to me, and although he did not speak the words, he lived them. I am incredibly lucky to have been his student, on all his adventures, even though, the consequences of his romantic actions had stiff results—like tin legs.

Bu's Lesson—Live as if you've never experienced any one thing before, and if you get stiff afterwards – it's easy to persuade a loved one to rub the sore bits out.

Chapter Twenty-Six
Clean-Up Crew

When I lived in Fremantle, for one reason or another, people used to dispose of their garbage in three of the lovely places where we walked in and around the port city. The oval nearest to where we lived suitable for a small walk was often covered in litter from the nearby college students. The medium lake walk, where people took picnics, had garbage caught in the edges of the reeds and melaleuca trees. The longer hill walks were dotted with tipped furniture and rubbish piles with dangerous content like broken glass. It was very sad to see and my heart felt heavy every time I walked in and around those areas. The litter from the college near the oval was so bad at one point that wrappers from the students' favourite fare of burgers would drift up my driveway. I joked with staunch vegan Tarryn that he needed to hide his secret much better; humour being the way to address my distress.

For a few months, I felt some degree of helplessness as I worried about how to solve the problem. Then, one day, coming out of a morning meditation, I remembered firstly the Buddhist proverb:

"If you have a problem that can be fixed, then there is no need to worry. If you have a problem that cannot be fixed, then there is no need to worry."

And then secondly, some wise words of Jiddu Krishnamurti, an Indian philosopher:

"If we can really understand the problem, the answer will come out of it, because the answer is not separate from the problem."

I set about understanding the cause of the problem, then decided to be at cause, to take action, deciding that the inertia and inaction on my, and everyone's behalf, was a part of the problem.

I began our twice daily walks with recycled bags attached to Bu's lead and then, looking at the volume of the garbage left lying around, I had a doggy backpack for him, stuffed full of larger garbage bags. As Bu and I loved to be out walking and in nature, the problem had inspired in me the urge to use the time to enact good citizenship. I called the relevant local bodies to inform them of what I was seeing and requested help. No help was offered. I was shocked at the apathy and lack of care of the natural places under their stewardship. The apathy and inertia were rife and it simply made my resolution stronger.

Bu and I kept walking and I kept filling up the bags and disposing of them in my own rubbish bins. After doing this for a few months, I noticed something incredible; other people were taking their dogs for a walk and picking up garbage too, using gloves in the case where the objects were dangerous. In a year, I noticed that none of the regular walkers took garbage bags along on walks anymore, the need had gone. The absence of rubbish was causing no or significantly less rubbish. A fundamental shift had occurred at both the oval and the lake. The answer was nested in the problem. All it needed was a casual clean-up crew of willing volunteers and helpful hounds, none of it coordinated in any way.

The dumped couches, car bodies and household goods in the hill walk took a more concerted effort of lobbying local councils to clean these up and return the bush to its pristine state. Sadly, not even two years after the clean-up, the place was littered again with much the same thing. As I had moved into the Perth Hills, I merely sighed, Bu trotting at my side and looking happy to be marking his old territory, and I took my energy into new projects closer to home.

There is a lovely long walk along Fremantle Beach groyne that Bu and I often enjoyed in the evenings but it too was littered, not with garbage but dog droppings dotted at every couple of feet. The smell was not pleasant at all, even in the cool evening sea breeze. Bu and I deployed the backpack and the bags once more, as did Tarryn, picking up kilos of the stuff on each walk. In a short time, we had made a difference and similar to the oval and the lake we frequented, there was no more dog poop.

I noted with some satisfaction recently, that the council now has scoop bags at the entrance of the groyne, and they are being used. These days, the walk with my new dogs along this wonderful structure, that gives an unprecedented view of beaches and ocean, is only scented with salty sea air. It just took a small and dedicated clean-up crew to start making inroads and seeing the solution in the problem.

Bu's Lesson—Take your devotion out for a walk (every day) with a bag or two and make a real difference. Be the solution – that elusive antidote to what ails you and your surrounds.

Chapter Twenty-Seven
Analogue Tracking Device

Bu loved to roam and most of the time, he stayed within my view, but the lure of rabbits, kangaroos and those tasty poo treats were too much for him at times and he would skip away as soon as he could. It would only take a moment of my distraction, like stealing a kiss in the cold air with Tarryn, admiring a spring orchid on the bush floor or just looking out at a view. His sense for me not having my eye completely on him was phenomenal.

Bu caused moments of terror for Tarryn, who stated to fellow dog walkers when he went missing, that it would be wiser for him to emigrate to a distant land than return without Bu. He was a wise man—forgiveness would have been very difficult indeed. The only place this really caused any concern was the first time we went to an unknown location and I did not understand the landscape.

I told Tarryn shortly before taking off for a picnic and a long walk that the location we were heading to was, as yet, unexplored for Bu, and he might decide to take off and we'd have no way of telling where he was. Tarryn dived into his sea container—that which held his worldy possessions, placed next to the caravan on the farm—and with a sly grin on his face, popped out with a solution held behind his back. "*I have an analogue tracking device*," he said laughing. How I loved his laugh, it was infectious and born of a true, deep-down happiness that I have witnessed in very few people.

We drove out to some spectacular state managed forest and sat down to enjoy our picnic, Bu on a lead by our sides with his own snacks to consume. We dawdled in the soft sun, chatting and holding our camping mugs of delicious spiced chai tea. Bu drowsed on his blanket, content with the sun warming his back and legs, twitching gently, chasing something in a dream. I can still feel the sheer luxury of the scene, the quiet hush of the bush, the peace within and the feeling of contentment. As it grew cold, we decided to warm up with a walk and explore the area which had a little winter creek criss-crossing the land.

Tarryn grinned up at me and said to wait for the deployment of the analogue tracking device and waking Bu up, led him away to the car to do just that. I closed my eyes, waiting for the surprise and could hear something metallic. My suspicions grew. Tarryn lived on a farm, with cows, there was a muffled clanking sound, and sure enough, there was Bu, trotting towards me with an enormous cowbell attached to his collar.

It was way too big for him and reached down to his knees, so when he went to sniff at something, it dragged on the ground. We could hear his progress very clearly as he made his way through the bush, but within twenty minutes, he returned to our sides, looking decidedly grumpy.

He parked his furry rear end down in protest and looked up pleading to get the thing off him. How could we not give in? His pride was wounded and his progress and enjoyment of the moment hampered. It was a sad state for a grey muzzled beloved. I realised I had deprived him of his freedom.

Zen thinking suggests that true freedom is only in the 'here and now' and we can abide there in the emptiness of expectation. In the absence of lack we may feel ourselves as a pure presence, a unique consciousness flowing in and out of unicity. In trying to keep Bu safe and allay my own fears, I had disrupted this potential for freedom for all three of us.

So, we removed the analogue tracking device, and a much lighter and liberated Bu set off into the bush, coming back every few minutes to make sure we had not gotten lost. Giving Bu free play not only unleashed him, but it flowed back to us in the form of a creative impulse we knew inherently to be in our hearts, manifesting in an evening of joyful music.

Bu's Lesson—Freedom is always available to you; just unbuckle whatever weight it is you are carrying around unnecessarily – and put it down out of reach so you can't pick it up again.

Chapter Twenty-Eight
Stairway to Heaven

You know already that Bu loved all of his many beds. In my Fremantle house, he did not have access to his most favoured one, the guest bed, because it was located not only in my art filled yoga studio, separated from the main house, but it was up a very steep ladder that even humans had trouble clambering up and down.

Guests loved the studio with the light filtering in through the stained-glass window and the utter seclusion of the private garden with its pink flowering princess gum tree. They loved the peace of it, but they all looked upon the sleeping mezzanine with some degree of trepidation. It was up a very steep ladder!

If I was working on one of my art projects, Bu would frequent the chair placed in the corner of the studio that was just for him and snore away peacefully, 'BuBu, BuBu, BuBu'.

If I was doing yoga, the little scamp would grow bored in his corner chair and try to join in by demonstrating what downward facing dog actually looked like, as opposed to what I was doing. Bu would then execute the frog pose, legs splayed behind him with his heavy head resting between his front paws, sideways across the mat.

"*Very helpful,*" I would tell him, "*the demonstration, not the parking yourself in the middle of my mat.*" I took it as a sign that enough yoga had been done and I was to follow his example and find a position for meditation. He was the Zen master in this. He showed me most days to stop doing and focus on being, letting the agitated mind come to calm and the restless body come to stillness.

He sometimes used the opportunity, as I sat in meditation, to pace around the room and create mild distractions for me to be aware of but not pay attention to. Zen Buddhists learn this 'art of mindfulness' exercise as one of the practices of the Eightfold Path. The Noble Eightfold Path is an early summary of the path of Buddhist practices leading to liberation from samsara, the painful cycle of rebirth. The meditator is taught to let noises, sensations and thoughts filter through the senses, acknowledge them and then let them go. I couldn't help myself one morning though, there was a distinct scrabbling noise of claw on wood and a gentle 'ooff' (yet again) of a furry tank trying to heave himself up somewhere.

Opening one eye, lest my Zen master notice my inattention, I spotted him with his front paws on the lower rungs of the ladder going up to the mezzanine and his right back leg valiantly trying to find purchase on the lowest rung to make the climb. I think he would have tried a version of his palm shimmy, but the ladder struts were just too wide apart for that particular manoeuvre. He glanced back after a couple of attempts to see me silently laughing, and after eyeballing me with the equivalent of a Zen teacher giving an inattentive novice a whack on the shoulder with a 'kyosaku' stick, took himself away with as much aplomb as he could muster in the face of defeat.

He would rarely sleep while I worked in the studio on my projects and instead, he'd watch me patiently, shifting positions if he got uncomfortable from being curled up in a ball—the bed awaiting upstairs too taxing a task for one small, and by then, mildly arthritic dog to achieve.

To untangle his legs and move his spine after so much stillness, we would head off to the nearby school oval and take a spin—working out the kinks in both our legs. There was always a reward for a pooch's patience.

The guest bed being out of reach, each day Bu conquered the steep spiral wrought iron staircase in the main house mezzanine level and slept there. When I arrived home, he would labour his way down, one step at a time, his nails tippy-tapping out a metallic rhythm, stopping from time to time to sigh as if to tell me the set up was not to his liking at all and that his ageing hips and legs were almost too challenged.

Still, the joy of resting on my bed and angling his furry body into the shafts of filtered sunlight was too much temptation to overcome. So, he persevered climbing the stairway to doggy heaven; a soft bed, adorned by a fleece rug atop it, ready to be graced by the Zen master, Bu.

Bu's Lesson—A snoring, ladder climbing, yoga mat hogging dog can provide delightful distractions to help you pay attention to each moment – to let go of mind and simply be.

Chapter Twenty-Nine
Resonance Therapy

Bu was a very sensitive soul. Inside the body of a small tank, beat a sweet and soft heart, pulsing with wisdom, alert to everything. He reacted though to any form of energy healing like Huna or Reiki. He did not like them at all and showed his disturbance by running out of the room as soon as I or anyone else intended to offer some healing for his failing health. If I tried distance healing from another room, in he would run and puff excitedly at me, "*Jesus, Lady! Get out of my aura already*," he seemed to say.

At first, I thought it was just me, but a number of friends trained in other healing arts tried and got the same anguished breathing. Massage, on the other hand, he would take any amount of, as most dogs do. He would beg for more if he was not quite satisfied with the effort, I had put in by putting his paw up on my shoulder to indicate that the session was not over yet. I knew I had found the right places to ease his aching muscles by the happy 'BuBu, BuBu, BuBu' emanating quite loudly whilst I worked.

Huna and Reiki are ultimately mystical to me. I am really not sure how they work but I have seen profound healing in people after I've given them these treatments. It is a pleasure to see a stressed and aching being, human or canine, get up off the healing table, calm and alert as they float up the passageway, out the door and back into the real world. It's how I feel when I come out of a deep meditation, experiencing the quiet elation of dwelling in the heart of an informed practice, rather than borrowing something from the surface of life and gaining unmerited grace.

Bu, despite his dislike of receiving, was a generous soul and loved to give and was part of most of my healing sessions with clients. He would invariably follow me and my client into the healing room, where he would insinuate himself under the table and wait for the energy to begin to flow. Nobody ever refused him entry, it was too adorable seeing a dog crawl under the table and half close his eyes, as if saying a silent mantra. He looked inscrutable and venerable like a wise old monk.

Bu would begin to snore as he nodded off, vibrating the table gently from underneath. Clients mostly got used to the extra 'resonance therapy' treatment. If I ever mentioned it, they would just laugh and either say they had not noticed or that something special was added to their time there, which was somehow enriched by the canine contact. For a dog that really did not like to receive energetic healing, he certainly loved to be in its backwash from others.

This was a paradox to me. Zen Buddhism is full of seeming contradictions and paradoxes. Students learn their poetic language through solving koans or riddles. These are designed to throw off the shackles of brain based logical reasoning and immerse the mind into the nature of experiential being. A koan's best answers are learned through the heart.

Bu was my teacher, a living koan, an endless riddle to fathom daily with heart, evolving my Buddha nature as a true Bosatsu does. It is the ultimate gift of love. Bu's own resonance therapy was the action of snoring, but underneath it were waves of compassionate and selfless love for whomever had turned up to be with us at that moment in the healing room. My other loving dogs, Diesel, Stella and Jessie, have shown no interest whatsoever in becoming healing companions, although Jessie does love to jump onto the table and get a good rub!

I feel so lucky to have had my companion healer with me all those years, amplifying whatever was given in his own mystical fashion.

Bu's Lesson—Riddle the heart daily, love beings as they love to be loved – not how you love to love or be loved.

Chapter Thirty
A Few of My Favourite Things

Bu spoke to me daily, through his antics, misadventures, pressing up of ice-cold nose to warm skin and happy breathing 'BuBu, BuBu, BuBu'. He communicated with the brilliance of a Bosatsu to me. His words spoke in dogs' language of what he needed, wanted, desired, all of it sent with a profound wave of loyal love. It was the Shamanic discipline of Shinto that taught me how to listen to different creatures and none more clearly than canines. These are Bu's favourite things in his words…

My mum fed me really well. In winter for breakfast, after my walk, I had porridge with turmeric, ginger, dates and blueberries. It was simply scrumptious and if she ever left the pot on the stove and went to work without soaking it, well, let's just say I was always willing to lend a paw, apply tongue and it was washed by the time she got home. In summer, I had a cool quinoa cereal mix made with bone broth and bits of tender meat floating in it, absolute heaven to my nose.

It was worth getting out of my basket every day, going for the first walk of the day and on returning home, seeing what was on offer. After Mum had fed me, I made sure I went to her and said, "Thank you." I was grateful for the food; I had been starved at one time, so each meal meant a lot to me. Plus, I know if you say please and thank you, it generally attracts what you want sooner rather than later.

After my breakfast, I would cruise to wherever Mum would be sitting and turn on my cutest 'starved pup' look. You know the one—head to one side, jaw slightly down, eyes peeping up—and she succumbed every day, without fail.

Down her hand would come and I would get the glorious crunch of a corner of toast, with a hint of jam if I was lucky. If I looked really cute and endearing that morning, I could also entice her to take out some tidbits of my favourite thing.

My favourite thing was roast chicken, it drove me mad—the smell of it cooking. I felt the drool trickling down my muzzle as I waited for it to come out of the oven and be cooled. It's a shame I stole that chicken at her friend's house because after that, she would put it way out of reach and I would have to sit there on the kitchen rug, meditating hard on opening the pantry door, almost feeling delirious from the intoxicating smell. I would all but swoon when she fetched it out of that magic pantry cave, the one that had all the human goodies and my treats stored in it—the one with the handle fixed upside down that I simply could not tackle!

I got so excited when she peeled off the crispy skin and stripped the tender flesh away from the bones that I would leap up and down in anticipation, executing a happy dog dance. I am sure it looked stupid, but I couldn't stop it. I'd get beside myself and start that silly breathing I couldn't help doing when I was relaxed, just happy or in the case of roast chicken, excited, 'BuBu, BuBu, BuBu' I'd go—gee whiz, it was embarrassing.

Salmon and its charred skin were great too. I don't think I drooled that much but a piece of this on basmati rice was a good dinner to have. Mum would have some pile of weird looking green stuff which is strange because I am guessing, to bring home all that meat, chicken and salmon, she must be a great hunter and fisher. A better hunter than me I know—I never did get one of those annoying possums, frogs or even a rat from the palm trees, no matter how hard I tried. Still the chase was fun!

Boy did I love cheese, I can't understand why humans got so upset when I tried it—it was just sitting there, right at my head height, clearly laid out for me—several types, all on one plate to sample. What a feast it was too, with butter on the side! I was about to start in on the slices of bread, but they really kicked up a fuss, so I went outside to ferret out one of my 'for later ons' from a pot plant.

Anything was good for me, really, I wasn't fussy. I loved the homemade treats, especially those special chocolates made just for me from carob and the vegetable biscuits shaped like a bone. Mum really made sure I didn't eat bad things like onion, macadamias or grapes, all like poison to a dog.

I had a night though where I ate all the left overs on people's plates and lapped up all the weird tasting red water (who says we dogs don't see colours) when everyone else was sleeping and I felt sick, very sick. Sick as a dog in fact!

I thought I had been poisoned as I laid on the lawn, unable to move even the next morning until I got fed delicious, warm sausages and hash browns straight from the barbeque. They made me feel a bit better but I never ever touched party leftovers again. My mum liked having parties I think. I loved them too as there was always inevitably someone, I could give the sad look to who would feel sorry for me and then feed me tasty morsels. If the look didn't work, I could escalate my chance of food fall by crawling under the table and moving in between someone's legs and putting my head in their lap, my teeth suggestively on view, until they gave in.

Not yet calling me the master manipulator, err sorry, Zen master? Well, I could get Frank the Texan who painted my mum's house to feed me whole Dutch almond biscuits just by sitting right in the place where he needed to move his ladder. It was so much fun looking up at him with my big brown eyes and looking pathetic, sheeze—I can't believe he fell for it—every time!

I remember being very hungry after he left though, my food rations were cut severely. There was definitely less in the bowl than usual. Perhaps I had overdone it a touch begging for those biscuits. I was definitely having trouble getting on top of the compost bin for a while—my furry little butt did seem heavier somehow, trying to jump up for inspection. When my diet was over, Mum must have felt sorry for me, I got a quarter of a mouth-watering roast chicken. How I loved roast chicken!

What I did not love so much was when Mum or anyone else tried to do some weird energy thing and 'heal' me. It made me feel prickly and tickled. I'd escape as soon as I felt it. What I did like to do was insinuate myself under the healing table and gently let off gas (best in show aromatherapy) and fall asleep until I snored. (Mum didn't mention that bit before – she's too polite!) I felt that it helped make humans feel more comfortable, safe even, if the prickly and tickled feeling would start to get too much for them. I don't know how they came out looking so relaxed.

Walks spent hunting were fantastic and rolling in fragrant carcasses was a pleasure. Presenting Mum with dead treasures was simply hilarious as she tried to look incredibly pleased and not completely disgusted—which she clearly was, when I would lay down to feast in front of her. I am sure I saw her almost gag once when I used all my strength to bring back a whole kangaroo to show her how much I adored her. Well, it smelled good to me in any case and I can't think why we left in such a hurry. We walked some more, but I'd got tired after lugging that great big thing through the bush so I slept on my bed, which was actually Gary's couch, as soon as we got back. I liked Gary so much, when Mum was sleeping in (so rare for her), he would take me out into the bush for a ramble. He smelled natural all the time and he was definitely one of my favourite things.

I had beds everywhere, one outside in the sun, one outside in the shade, one basket in Mum's room, the guest bed (read chapter 17, It's My Bed!) and my sand pit. I could move around as I liked and enjoy every aspect of the day in complete comfort and get into the heart of those matters that require deep meditation and sometimes, dreaming. Matters like *"I wonder where Mum goes when she is not with me? I wonder if there is time to catch some more sleep? I wonder if there are any rats to see off the property?"*

I dreamt a lot, I'd wake sometimes with Mum's hand, warm on my flank, whispering, "*Are you okay*?" in my ear. Of course, I'd been twitching—why wouldn't I be?—chasing rabbits takes athleticism, even in dreams. At Bruce's farm, the one from where I stole the poo crackers and slept in the tipi, I needed first thing to put the cows back in order and show them who's boss. It was exhausting running back and forth.

I had a mat on the caravan deck to myself and looking out across the fields, I would grow drowsy in the sun and Mum would curl around me sometimes and I'd be warm and happy. Ever heard the expression 'dog days'? Mine were *great* dog days indeed. Sometimes, there were great dog nights too, gazing out across the bush as Mum looked up at the stars from the caravan deck and stroked my head. I loved being there as I loved and trusted Tarryn, he was amongst my favourite things—now *there* is a human who knows how to mooch properly. In fact, I'd say this Zen master was surpassed by his student!

My friend Adam, a dear friend of my mum's and a definite favourite of mine, adopted me as a mascot for his field of work. I didn't do much to earn the title except to show my complete disdain for all the yakking that went on when he and Mum sat on the couch. I endlessly showed them how to just be peaceful and quiet by curling up on one side of that beaten up, saggy piece of furniture and snoring loudly but they persisted on talking, so I just slept, as close to the heat as possible in winter, and in summer, nearest the breeze.

I liked being the mascot, it was an important role and I took it very seriously. They consulted me at times, but they didn't really listen to my answers carefully enough—I gather that humans are just dumb dogs—I mean really, they can't smell that well, they are hard of hearing, they don't see clearly, if at all, and they spend so much time worrying about small stuff. They should sleep and dream more and let go of the endless suffering that worry causes. It's all in their heads, speaking of which, what's with the fur only on their heads? Maybe that's why they don't see life clearly?! They work too much and are out of harmony with themselves and others. It all just looked dumb to me. As mascot, I did my best but honestly, humans huh?!

Besides sleeping and dreaming, I loved to chew guests' thongs on occasions (fun to send them off with only one and a half left to wear), bits of wood inside the house (fun to see humans hopping around, clutching their weird shaped toes with bits of it stuck into them), rope toys (fun to tug really hard and then suddenly release so humans fall backwards on their furless bottoms) and a ball that made some horrible noise that really bent me out of shape. It would take me just minutes to get rid of the annoying ringing and then, I could spend hours gnawing on the ball. I loved that ball without the ringing, but no matter how much effort I put in to destroy it, a new one popped up in its place—life was amazing.

Talk about amazing, those potato crisps were scrumptious but I only got to try them once in the back seat of Mum's car as we were driving to Nannup so I could impersonate some extinct trumped up dog thing, the Nannup tiger. She explained halfway through the packet, when she realised what I was crunching on, that it was the apple she'd given me, not the crisps. Well, you can't blame me—a mere piece of fruit is not exactly a first-choice food for a dog except perhaps for watermelon.

Grandpa used to feed me watermelon that was so sweet and so good. He would chop it into small chunks and we would sit on the couch together and eat a huge bowl of the stuff between us. I loved Grandpa, he was always kind to me even when I got a little out of control once and destroyed a metal doorknob, eating my way out of his study to be on the other side of the door with him. I missed being next to him, that was all.

You know why I wanted to be near him—he was one of those rare humans besides Mum who could actually, really and truly, hear what I had to say.

Sometimes, after visiting Grandpa, who lived near the coast, Mum and I would jump in the car and head to the beach to share fish and chips. She ate the chips (I tried one too, but seriously—yuk, what horrible things) and I had the sweet and yummy fish. I thought we made the perfect pair, sitting at the beach, sharing food.

I loved leaning into her side, looking out at the ocean as we ate; just me and her, endlessly loving one another in that moment of quiet calm. When we had finished eating, I would rest my heavy head on her shoulder and look out to sea, trying to see what she was seeing. We would breathe in time and with me pressed close, I noticed that she relaxed (almost as well as I did) and was completely herself.

I wished I could have stayed with her forever. I left with so many lessons untaught like how to mooch for hours (not just minutes), how to enjoy the things that matter most, to rest deeply in the sun and cool off afterwards, to snuffle the intrigue of the traces of scent on grass, to ensure being stroked until warm on a cold day, to delight in a walk through the bush, and, I can't say this enough, to savour the sweet anticipation of eating roast chicken!

Bu's Lesson—Surround yourself with your favourite things and people—you will find someone on whom you can rest your heavy head who will love you in each and every moment. And if you're lucky, that love will come with a side of roast chicken.

Soulful Bu

Art credit to: Nada Orlic

Chapter Thirty-One
Saying Goodbye

Bu was born on the 17th of July 1999 and was four-and-half-years-old when I brought him home for good on the 23rd of November 2003. It still came as a shock, despite all the illnesses and injuries and looking at the greying muzzle each day, when Bu spoke to me very clearly of leaving. Hadn't I only just picked him up from the refuge? Weren't there so many more adventures to be had, more roast chicken to eat?

I was sitting in the lounge room, cross-legged on the floor listening to music and wondering where Bu was when he came up behind me, circled around and leaned his whole-body weight against me and plopped himself onto my lap—a thing he had never done before. He stayed there for about three minutes and let loose a wave of anguish, hidden from me for some time, I now know. He made it very clear that it was time for him to leave, that he did not want to bear the pain of his body anymore and the terrible dragging tiredness that was making each day wretched. He looked up at me, into my eyes the whole time, he sat there, cradled in my lap and I understood perfectly.

The ultimate compassion was being requested of me and I had to show up and end the suffering. After the request had been made, in typical Bu fashion, he got up and jauntily exited the room in search of a sunny spot on a cold day to warm his aching bones.

It took me two weeks to action Bu's request. I had to find the 'right time' which it could never be for me. Every moment afterwards seemed drenched in sorrow, and each time Bu and I walked, it was with greater awareness than ever. For me, it was like I had never seen the bush so alive and vibrant nor Bu so free-spirited as he ambled along—arthritic hips clicking—barely able to see, but content enough to smell, and a poo connoisseur to the end, he sampled a few kangaroo droppings.

I made Bu all his favourite dishes, roast chicken and crispy salmon skin featuring heavily. He only ate a little but still pressed his nose against me to say thank you after each meal.

In despair at the imminent loss, I finally called the vet for a home visit and then I called Tarryn who immediately flew back from his overseas trip, cutting it short to be with Bu and me. We went for what turned out to be an incredibly vigorous walk on the afternoon before Tarryn returned.

I could not believe how fast we went; I yearned to get home and cancel the vet visit. Bu collapsed in a happy and panting heap 'BuBu, BuBu, BuBu' when we arrived home and putting a paw on my arm, reinforced that the walk was a last hurrah and he still needed to go.

The night before Bu's final passing, he slept sandwiched between Tarryn and me for a few hours on 'his' bed and then went outside to sleep under the stars for the remainder of the night—I missed snuggling, but he was independent to the very last night, the very last moment, and I admired his tenacity of spirit.

The day of his death dragged as a string of Bu fans visited, including my dad, who brought an extract of the 'Rubaiyat of Omar Khayyam' poem to read out and comfort me in losing my soulmate.

"Give me a flagon of red wine, a book of verses, a loaf of bread, and a little idleness. If with such store, I might sit by thy dear side in some lonely place, I should deem myself happier than a king in his kingdom."

Everyone bade him a fond farewell. I felt vacant and sick, spending most of my time on his bed, whispering how much I loved him again and again. I tried to meditate but peace was eluding me, so I just went with the turmoil and spent time recalling his funny antics, incidents and sharing the moments like a crazy filmstrip with Tarryn who wisely listened with the utmost compassion and made very little comment except to recall funny stories which are now, dear reader, all part of your knowing Bu.

The vets came in the late afternoon, Bu greeted them with his stumpy tail, *'thump, thump, thump'*—always just the three times—leaping about like nothing was wrong with him and that he had not made the request to go. I was sorely tempted to send them away but knowing that a prolonged suffering from kidney failure, arthritis, cancer and a whole stack of other issues was all he had to look forward to daily, I welcomed them into the house.

In a strange paradox, these was the very vets who had saved his life once from the near fatal sand and toxic buffet incidents and then as he grew old and ill, gave it ease and comfort in the three years afterwards.

Bu was given sedatives to calm him and give us time to speak to him, but these had an opposite reaction and he leapt around like a puppy. He trusted me as I called him to me and he was given more and eventually, he settled in my arms but asked to go outside to his bed, so we did. The two vets and Tarryn and I surrounded him, and speaking of love to Bu, the vet found a vein and administered the final relief.

Bu and I looked at each other one last time, an exchange of hearts about to be broken in parting, promising to find each other once more, then he closed his eyes and went to sleep with a last breath of 'BuBu, BuBu, BuBu'. Tarryn and I stayed with him until the kind and understanding vets got us up away from his eerily still, russet coloured brown and black striped body, and carried him to the car in a patchwork blanket. I got one last glimpse of him as he was put in the car, his head relaxed to one side, looking quiet as if in deep meditation.

The vets drove off. Tarryn and I went inside the house that now felt empty and bizarrely foreign and I sent out a huge pulse of love to wherever his spirit had gone, a kind of beacon to find me should he need anything or to find his way back. I then fell to my knees on the exact spot he reserved for mooching in the lounge room and sobbed and sobbed while Tarryn held me.

We visited the farm quite often afterwards, to walk and find peace in the wake of the sucking void that Bu's death had left and we both gathered serenity in the olive groves, listening to the cows cropping the grass and the chickens fussing around the place. Friends would express their love for him in chants inside the tipi at Full Moon Drumming, a human devotion to the divine that resided in his sweet soul.

In an extraordinary show of love, months after Bu died, Jemmy would still rush to the car to look for him and seemed to us, confused that he was not in the car with me. I found it incredibly challenging to know that this would happen every time I went there. Jemmy passed just a year later, and I hoped she would somehow feature again in Bu's next life, they were certainly sweet on one another.

Tarryn went off adventuring again when he had determined that I would be fine alone without both him and Bu. Discerning that exact moment showed a huge amount of grace. Returning from the trip early to say farewell, staying to be with me in the wake of the loss of Bu, and leaving just at the right time demonstrated such compassion and love. I felt humbled and honoured.

I still sobbed a little each day as I gradually divested the house of all Bu's beds, bowls and toys until just his collar remained alongside my favourite photograph of him. I felt his presence around the house and garden during the first few weeks of his passing, imagined I could hear his claws tippy-tapping down the passageway to go out to the garden or hear the swinging of his dog door flap as he went to snooze in the sun on the front deck. I kept a corner of toast each morning on the plate as a reminder of delicious things shared at the breakfast table.

The day I said goodbye to Bu was the day my heart rendered asunder. His death is something that I describe as a broken heart that will never quite heal. When I thought of him, and when my heart gripped in anguish of his absence, I recalled the Zen lesson of pulling in pain and sending out love; cultivating a noble heart to awaken Bodaishin. This simple, yet profound practice helped me find stillness and increased my capacity to love and be loved.

I found peace in returning to my meditation practice, quietly missing Bu's snoring in the background. I'd grown so used to it, but the absence no longer distressed me. I thought often of the noble heart contained within the shiny, red, brindle-coated Bu and how fiercely and fully a life can be led with just a little help and love.

Bu was fifteen years old when I said goodbye for good on the 14th of June 2014. We had ten and a half incredible years together and abandoning all Zen sensibility—because I am a frail human who loved without reservation—I still would do anything in my power to have ten and a half, no, a lifetime of years more.

Bu's Lesson—Listen to your noble heart, there is no maybe about it – we will always find the ones we love, each time we come around and love will continue to grow.

Chapter Thirty-Two
In My Dreams

In the weeks and months after Bu's passing, life did begin to feel as if joy could be found again. The small daily encounters with friends and their beloved pooches reminded me to look for grace in the ordinary and count myself extraordinarily fortunate to have experienced a most pure and innocent love for over a decade.

One month after Bu died, I dreamt I was walking alone in a park, my head bowed down by grief, wondering if I should sit down and meditate on the grass and fill my lungs with sweet bush air and my heart with the love that still pervaded my whole being for Bu.

In the dream, I saw him, somehow finding me in this park, that was unknown to us both. We walked alongside each other, me in bliss and he just snuffling the ground until we came to a house on the edge of the trees that defined the park. Bu threaded his way through the trees and into the house. I followed him inside and stroked him, enjoying the feeling of the velvety softness of his head under my hand. He leant against me briefly, walked back towards the door and dissolved through it. I ran, terrified at losing him again, opened the door but could not follow, an invisible force field preventing me from trailing after my furry beloved. Bu looked back just once before disappearing into the trees.

I woke refreshed for the first time in four weeks, feeling he was somehow safe and was moving on towards his next great adventure, and that he had left a message that it was time for me to move on too.

One year after finding peace (more or less) with Bu's passing, I had another dream in which I went to fetch a dog from a refuge for a friend. Instead of going in and locating the dog, I saw Bu in a kennel and knew that he had been there, very sick and old for a year, all the time waiting for me. I was horrified as I thought he had died but hadn't—there is no logic in dreams sometimes. I got him out of the kennel as quickly as I could but in gathering him in my arms, he dissolved.

I woke up shaking and sick, looking for my car keys to go and get him but the insanity of it was realised soon and it gave me an opportunity to sit still and once again honour the time I was blessed to spend with a dog who was once a refuge savage, now a furry angel.

Three years after his death, I dreamt Bu was walking up the street, the very same one that Tarryn and I had once caught him out on. As he walked along, he began to rise, a gentle wind seemed to carry him into the treetops. He continued onwards and upwards, blown by the wind, until he sailed up into the clouds with a sweet little doggy grin, his eyes shining liquid love.

He looked down at me, floating free and happy in the swirl of dancing leaves and I knew that he was utterly content, in an ultimate state of Zen, wherever he was or whomever he had become...

Bu's Lesson—Let me float free and dance as a speck of dust in the wind for a while. You know I will return and once more be with you – how could this perfect love not draw us near once more?

Chapter Thirty-Three
Three Anniversaries

Bu's ashes were scattered in two of his favourite places just one week after he died. Half are in the serene Nannup bushland where he performed thylacine duties and the other half are strewn across the granite rocks just down from where he slept in the sun on the deck of the house, and where he passed his final hours, mooching in the sunshine—a Zen master to the last.

Tarryn and I scattered the ashes together with remembrances of Bu spoken softly in the chilly June air. Tarryn played the didgeridoo in a hauntingly beautiful honouring of Bu's extraordinary power to embody the truths of not only Zen but also Shinto and Huna. I accompanied Tarryn with a chant of Japanese Shigin poetry, which while not strictly Zen, mimics the philosophy closely enough. We wanted to send Bu off in style, as befitted a dog with a deep soul, the bold adventurer and the instigator of merry mayhem. Five anniversaries have marked Bu's eventful life since then.

On the first anniversary, I visited Bu's favourite walking places around Perth and Fremantle—he had lived in so many places, it was difficult to cover them in one day. The memories came flooding back and I began to jot these down—not without a tear or two in the beginning.

Just a couple of days later, I went to Nannup. I held a solemn ceremony with Gary in the bush where Bu's ashes are scattered. I chanted again, hearing the echoes of Tarryn's didgeridoo playing in my mind. I felt my heart swell in infinite love and remembrance of all that Bu had taught me and although it brought his absence of companionship very close, I did not weep.

The site of his ashes felt peaceful, well it should have been—there was no Komatsu to disturb the tranquillity! The two of us recalled his misadventures with laughter and I felt a little bit more sorrow dissipate. As we talked, I realised just how many incidents and accidents Bu and I had been through together. The idea of a book came to mind on this day, if only for helping heal the space I felt that his absence had left.

Finally, in that first year without Bu, a dear school friend of mine and fine artist, JJ, offered to send me a gift after I sent a copy of my favourite photo of Bu to him. I was not sure what he wanted it for, but I didn't have to wonder long. We'd spoken often of our beloved dogs and I had seen an amazing painting of his old black Labrador.

I should have guessed. In the mail, a couple of months later, came an unexpected package, which when unwrapped, revealed an incredibly detailed and lifelike painting of Bu. JJ had captured him perfectly on raw canvas. It hangs in my study and I look at it most days with a smile, not just for the image of Bu but for the thoughtful and loving gesture from my friend that tells me in colour, how much he cares. JJ had been a steadfast and true friend through the years, transforming from a wild boy of curly long black hair—always the suffering artist, to a grown man full of the creative spirit but calmer now somehow.

On the second anniversary, I headed once more to Nannup, but this time, Gary was away camping in the Simpson Desert and Tarryn and I had become just friends. I went with my handsome and beloved Fuzzy Bear and my new dogs, Diesel, Jessie and Stella.

We visited Bu's special place in the bush that he loved so much, there was still a feeling of the utmost peace, broken only by three dogs bounding around the place excitedly yapping and doing what dogs do—marking their territory—no respect for the dead!

I repeated the chants as I had done before, lighting incense and then sitting still on a huge fallen Jarrah tree log that marks Bu's resting place, and I talked to him as if he was still right there by my side. I got a distinct feeling that he had truly moved on and yet, there was a small waft of breeze that played by my side. I felt happiness radiate upwards from the forest floor and move towards me. Looking down, I thought I saw the outline of a small Nannup tiger, liquid brown eyes, looking up lovingly and patiently as if to say—*"You are still attached? Wow, slow learner aren't you, let me go Mum, let me go…"*—so I did as the Zen master commanded, and I felt better than I had in the past two years. I could finally take an unforced deep breath and exhale with ease and sit in quiet meditation.

On the third anniversary, I paid what I think was the final visit to Bu's resting place. Diesel, Stella and Jessie were dropped off at a luxury dog retreat, so they did not repeat any gravesite desecration. There was no chanting, no incense, no leaving of smoked salmon—I just took a long walk—following Bu's preferred path, right to his favourite turn around point.

Heading back, I felt returned to a state of grace and able to tell Bu's remarkable stories. My dear partner, my Fuzzy Bear and I headed to our own retreat (not quite as nice as the dogs' one, I have to say) and I began to write, '*The Book of Bu – Tails of a Zen Dog*'. A book dedicated to Bu and his many teachings.

As I wrote, and the stories unfolded, I knew too that this is a dedication to all the rescuers of canines and all the rescued canines everywhere. All around the world, many more remarkable stories are unfolding, and the rescuers are becoming the rescued.

Thank you, Bu, you have been my teacher, my companion, my furry beloved and *my rescuer.*

Bu's Lesson—Celebrate the together-life that was and let go into the lessons it brought you – and then just let go, for you are rescued.

Puppy Bu
First Day
Nannup Bush
Perth Hills Cottage
Bruce's Farm
Tarryn & Bu
Its My Bed
Fremantle House
Meditating Bu

Book Two

Chapter One
Gromit's Tail

My former husband, Todd, has at last count, nine rescue dogs living with him and his wife in Oklahoma, in the United States of America. When I sent a couple of chapters to him to read and make comment on, he sent me back a beautiful story of love and forgiveness—this is the first one of many all true 'tails' in Book Two.
"More Tails of More Zen Dogs". Todd's story of Gromit demonstrates the most gracious dedication to embodying compassion; he uses all his resources and heart capacity to care about the canines that show up in need of a home and a healing of the wounds they carry when they land on his acres.

Todd writes: I live close to what must be a dump dog site. I do not know if the dumping of unwanted dogs is a practice in other countries, but it is in Oklahoma. The dog develops a problem or does not behave like the owner wanted or the puppy gets too big and boisterous or whatever other thing the owner considers intolerable. The right and proper thing would be to re-home the dog. Instead, the person of low character will take that dog someplace in the country, put it out of the car and drive away.

All my dogs were issued to me by God that way. I am out doing things on my property and notice a dog sitting there or peering through the bushes. Holstein the dog took over a month to convince that it was safe to be here. Dexter, Kanza and Private Ryan, the dogs all just ambled up as if they knew this place was home. I have found homes for others. I was an infantryman and sometimes at night, after I left the service, ghosts gave me problems. Dogs keep the ghosts away, and I cannot imagine how I might be without them.

I was feeding my pack of six one night and noticed, outside the fence, down by my barn, a furtive black ghost skulking just out of the light. It could have been a coyote, a big possum or a fast raccoon, but I was reasonably certain a new dump dog had arrived. I took a bucket of water and a bowl of food down, set it out and went to bed.

The next morning, the dog was simply sitting by the empty bowl, which caused my entire pack to go crazy barking at him. I went through the gate, and there was this old Labrador awaiting some kindness. He had mange and was almost hairless on some parts of his abused body. His belly and backside were covered in sores and ticks, he had these hairless pads on the back of his bum, and his teeth were worn in a strange hourglass shape. *Oh, man, what had happened to you?*

He was so patient. His lovely deep brown eyes caused me to think of Wallace the Inventor and his dog Gromit, and there was his new name. I got some oatmeal shampoo and a bucket and set about cleaning the top layer or two of caked dirt and scabs. He had no problem with the indignity of getting such treatment on his first day. After he was dry, I loaded him into my truck and took him to the vet for an assessment.

Gromit might not have been, in that moment, much longer for the planet. He was in bad shape. I did not know if he was too pulled down to be saveable and had steeled myself for an unhappy outcome. It was morning time, and I had already decided that if he had to walk the Green Mile today, I would take him back out for the best day a dog could have. We would go get hamburgers, find a pond to swim in and a shady tree to take a nap under, and then, I would take him back at the clinic's closing hour for the sentence to be carried out.

Doc Simpson was a new veterinarian, fresh out of the university. Gromit was one of his first cases. After poking, prodding, taking scrapes and samples, Doc believed Gromit had a shot at getting better and so, we decided to give the old lad a try. The Doc concluded that he had likely been a sire at a puppy mill, and the hairless swollen pads on his backside were from years of having been rubbed raw on the back of whatever tiny cage he had been confined in.

The weird shape of his teeth was attributed to his gnawing on a wire door trying to get out. He gave the dog some shots, me a mange liniment to rub on him, some antibiotics, and some shampoo that wouldn't dry his skin out too much.

He could not come through the wire and be with the pack at first, as mange is quite transmittable. For a month at first, twice a day, I would massage in the medicine, then wash him. It was over an hour a session those first few days, working ticks, scabs and lesions out of his skin and getting the lotion into every crack and wrinkle. The sulphur in the mange ointment left Gromit with a peculiar aroma, not unpleasant at all, as one might expect, but a warm lanolin-ish scent I liked.

He bore all this with good dog humour and patience. Despite his gentle disposition, I could tell he did not like being stuck in the outbuilding. I loved him up quite a bit and slept down with him once or twice, though the pack was not happy to have me out there with some stranger. Later, to get my pack acculturated to him, we would go sit by the wire so they could smell each other and do whatever it is dogs do when a new one shows up.

Gromit healed well, his skin cleared and began growing new hair, and he put on weight. After a month and another couple of trips to the vet, Gromit was pronounced fit enough to come through the fence and mainstream with the other dogs.

At some point in this process, my lovely girlfriend, with whom I was in negotiation for what a marriage might look like, let me know that, in her opinion, if we were married, nine dogs might be one too many. I had six at that point, plus Gromit possibly and she had two more. My sister Cass, a rural mail carrier who lives north of me, a hundred miles or so, had a dog named Waffles, who needed a buddy. The solution seemed self-evident.

So, a couple of months after getting Gromit completely well, I loaded him, my girlfriend and my dad up and took him to my sister's place. My sister loves animals as much as I do, and I knew he would be loved and cherished. We drove home thinking that all was well.

Cass called me up a couple of hours later. *"Did he have any issues going to the bathroom? We are on a walk, and he keeps trying to poop, but cannot."* I informed her that I had never actually seen him poop, but had, over the weeks, checked several of his piles for worms and to the best of my knowledge, he had no problems going. It was a head-scratcher, but not for long.

Todd's rescue dog, Grommit.

The next day, on her way to work, Cass took Gromit to her new vet. Oddly enough, a young lady who had been a classmate of Doc Simpson's. She dropped him off for a check-up and diagnosis while she went to deliver mail.

Gromit the dog was all full of cancer. His prostate had swollen to the point his colon was obstructed and though the doctor was able to evacuate him, he would not be able to go on his own ever again. Cass called me with the news, and I went to get a second opinion from Doc Simpson. Only one short day ago, I had stopped into my dog clinic and had coffee with the young vet who was happy to have had a success story at the start of his career. In disbelief, he called the vet up north, they talked vet talk, and the sad verdict was pronounced.

I got my girlfriend and father back in the truck and we returned to the town where my sister lives. I went to the clinic, talked to the lady vet and gave her my sister's address and a hundred-dollar bill. Would she please come out there with her euthanasia kit after her workday was done? This good dog would not walk the Green Mile on some cold steel table.

We took him out of the office and to the store. I got a couple of tasty cheeseburgers for him. We returned to my sister's place and I took him for a long, last, sad walk in the fields. We looked at the horses. We pawed at a couple of holes. We sat in the cold autumn sun as it dropped to the west and I held him, my nose buried in his wiry fur to smell that lanolin sulphur smell and crying like a baby, the big tough Marine. I heard a car arrive, stood and walked that good dog to a little hill, sat down and he cuddled back into my shoulder. With my family in a circle around us, the vet gave him the shot and that good dog fell asleep forever in my lap.

I have his collar, but the Gromit smell is fading from it as time goes by. For some reason, I know I will grieve all over again when I try to scent him but can no longer. This story is hard to write. He is not the first of my dogs to go, and he will not be the last, but his loss is for some reason the most poignant.

Reflecting on this profound finding and losing in such a short space of time, I am reminded of a key insight of Zen into impermanence. As humans, we need to cultivate the mind to be equanimous in the midst of change and how we respond to it; embracing the dreadful transience of a precious life like Gromit's, like Todd's, like yours and mine.

We can find ease and grace in the shifting landscapes of our lives that change brings. We can let go the clinging, the sheer futility of it and find a space to feel the happiness of holding gently and being prepared to let go.

Todd, the muscled, tough Marine knows that by watching each moment arise and then letting it fall away, he may reside in peace and in the now. He still feels everything profoundly but does not let it weigh him down into despair.

Gromit's Lesson—Today's love is enough for me. No matter how much you want me to, I am not going to last as long as you will, so love me today.

Terminology & References
Some Zen, Buddhist, Shinto and Huna Terminology

Bosatsu (Sanskrit – **Boddhisatva**)
An awakened or enlightened being who renounces the experience of nirvana and is dedicated to remaining with unenlightened beings and to work for the liberation of all. Also, a person who has generated spontaneous bodaishin or bodhichitta but who has not yet become a Buddha.

Bodaishin (Sanskrit – **Bodhichitta**)
A term for 'mind of enlightenment'. Generally speaking, Bodaishin refers to a being that is motivated by great compassion to seek enlightenment not for themselves but for the benefit of all living beings.

Buddha
The Buddha is the founder of the Buddhist philosophy or can refer to a being who has completely abandoned all delusions and is liberated.

Buddhahood
The state of achieving enlightenment.

Buddhism (Mahayana)
A Sanskrit term for 'Great Vehicle', the spiritual path to great enlightenment. The Mahayana goal is to attain Buddhahood for the benefit of all sentient beings by completely abandoning delusions.

Buddhism (Zen)
Zen is a Japanese school of Mahayana Buddhism which emphasises the value of meditation and intuition rather than ritual worship or study of sutra.

Busshō (Buddha Nature)
Refers to the notion that the luminous mind of the Buddha is inherently present in every sentient being.

Buddha Seed
The root mind of a sentient being, and its ultimate nature.

Eightfold Path
The Eightfold Path is an early summary of the path of Buddhist practices leading to liberation from samsara, the painful cycle of rebirth. The Eightfold Path consists of eight practices: right view, right intention, right speech, right conduct, right livelihood, right effort, right mindfulness and right samadhi (meditative absorption or union).

Emptiness
In Zen, emptiness is utter inner silence; a quiescent mind, empty of the belief that "I" is a separate entity with an independent existence, apart from all others. Zen Buddhists discover the empty nature of the world primarily through meditation.

Enlightenment (Satori)
An omniscient wisdom achieved by seeing one's own mind directly without passing through brain intellect.

Five Precepts
Most lay Buddhists live by these and refrain from:

- Harming living beings
- Taking what is not given
- Sexual misconduct
- Lying or gossiping
- Taking intoxicants

Four Awakened Qualities
Most Buddhists strive to have these at front of mind and act when opportunities present:

- Loving kindness
- Compassion
- Joy in the joy of others
- Equanimity

Four Truths
The four noble truths were Buddha's motivation for leaving his home and taking up a spiritual life, to understand duhkha (suffering) and find a solution to suffering. The four noble truths are the answer that came to the Buddha as part of his enlightenment. These are:

- Suffering is all around us; it is a part of life.
- The cause of suffering is craving and attachment.
- There is a way out; craving can be ended and thus suffering can be ended.
- The way to end craving is the Eightfold Path.

Huna
Huna means 'secret' in Hawaiian. Huna is the healing art, earth science and spiritual shamanism of ancient Hawai'i.

Inga (Sanskrit **Karma**)
Considered to be the sum of a person's actions in this current state and in previous states of existence, viewed as predicting of fate in future existences.

Koan
A koan is a riddle or puzzle that Zen practitioners use during meditation to help unravel greater truths about the world and to discover themselves.

Liberation
The recognition that one's body, thoughts and feelings are in a state of constant change and hence the identity of 'I'. One no longer experiences 'I' as the focus, which is the cause of all suffering.

Mantra
A mantra is a sequence of words or syllables that are chanted, usually repetitively, as part of Buddhist practice. An example of a mantra is 'om mani padme hum'.

Meditation
The practice of Zen meditation or Zazen is at the heart of the Zen experience.

Mindfulness
Awareness, remembering that all things are interrelated, living in the present moment. It would be difficult to overemphasise the importance of mindfulness in Zen and Buddhism.

Nirvana
Sanskrit word for 'liberation'. Complete freedom from samsara and its cause, the delusions.

Reincarnation
Not mentioned in the book but rather referred to as 'come around again'—the belief that a soul will be reborn in another body.

Root Mind
The very subtle mind located at the centre of the heart. It is known as the 'root mind' because all other minds are born and die in it.

Satori
A very deep state of meditation in which notions of duality, self and indeed all concepts drop away. Profound satori is very close to an enlightenment experience.

Seishi (Sanskrit Samsara)
The cycle of uncontrolled death and rebirth or cyclic existence.

Shala
Shala is a Sanskrit word that translates to 'home'. A yoga shala is a home where yoga is shared and experienced.

Shoshin
Zen word that translates to "beginner's mind". An attitude cultivated by Zen practitioners of having an open, eager, and non-preconceived idea when studying a subject.

Sōgyō (Sanskrit Sangha)
According to the Vinaya tradition, any community of four or more fully ordained monks or nuns. In general, ordained or lay people who take Bosatsu vows or Tantric vows can also be said to be Sangha.

Sutra
A Buddhist canon or scripture regarded as having been spoken by the Buddha.

Three Poisons
The Three Poisons are lobha, dvesha and moha, Sanskrit words usually translated as 'greed', 'hate' and 'ignorance'.

True Nature
Same as Buddha Nature.

Yakushi Norai (Buddha Medicine)
A devotional cult in Japan worshipping the Yakushi-rurikō the Medicine Master of Lapis Lazuli Radiance from the Heian period.

Zazen
'Total awareness in an upright posture' or seated meditation. Unlike meditation done in some other spiritual traditions, zazen is practiced by being aware of the breath's passage, it neither involves concentrating on one subject nor on aiming to block out all thought.

Influential Books
Bec's Favourite Books

Bayda, Ezra, '*At Home in Muddy Water*', Shambhala Publications Inc, Boston, 2004

Benoit, Hubert, '*Zen and the Psychology of Transformation: The Supreme Doctrine*', Inner Traditions Bear and Company, Rochester, 1999

Cameron, W. Bruce, '*A Dog's Purpose*', Pan MacMillan, 2012

Chödrön, Pema, '*When Things Fall Apart*', HarperCollins Publishers, London, 2009

Grogan, John, '*Marley and Me*', HarperCollins Publishers Inc, New York, 2011

Harvey, Andrew '*The Hope: A Guide to Sacred Activism*', Hay House Inc, Carlsbad, 2009

Hanh, Thich Nhat, '*How to Love*', Parallax Press, Berkeley, 2015

Hanh, Thich Nhat, '*True Love: A Practice for Awakening the Heart*', Shambhala Publications Inc, Boston, 2006

Hanh, Thich Nhat, '*No Mud, No Lotus: The Art of Transforming Suffering*', Parallax Press, Berkeley, 2015

Joko Beck, Charlotte *'Nothing Special: Living Zen'*, HarperCollins Publishers Inc., New York, 2011

Jung, Gustav Carl *'Synchronicity: An a-causal Connecting Principle'*, Princeton University Press, New Jersey, 2011 (Vol. 8. of the Collected Works of C. G. Jung)

Halpern, Sue, *'A Dog Walks into a Nursing Home'*, Tantor Media Inc, Old Saybrook, 2013

Kerasote, Ted, *'Merle's Door'*, Harvest Books, United Kingdom, 2008

Koontz, Dean, *'A Big Little Life: A memoir of a joyful dog named Trixie'*, Random House USA Inc, New York, 2011

Kornfield, Jack, *'The Wise Heart: A Guide to the Universal Teachings of Buddhist Psychology'*, Random House, New York, 2009

Magid, Barry, *'Nothing is Hidden: The Psychology of Zen Koans'*, Wisdom Publications, Somerville, 2013

Masson, Moussaieff Jefferey *'The Face on Your Plate'*, WW Norton & Co, New York, 2010

Mitchie, David, *'The Dali Lama's Cat'*, Hay House, London, 2012

Phelps, Norm *'The Great Compassion: Buddhism & Animal Rights'*, Lantern Books, New York 2004, p. 76.

Ricard, Matthieu, *'The Art of Meditation'*, Atlantic Books, London, 2011

Ryan, Tom, *'Following Atticus'*, Penguin Books Ltd, London, 2011

Stein, Garth, '*The Art of Racing in the Rain*', HarperCollins Publishers Inc, New York, 2009

Sumin, Haemin: '*The Things You Can Only See When You Slow Down*', Penguin Books Ltd, London, 2017

Suzuki, Daisetzu Teitaro, '*Introduction to Zen Buddhism*', Grove Press, New York, 1994

Suzuki, Shunryu, '*Zen Mind: Beginners Mind*', Shambhala Publications Inc, Boston, 2011

Trungpa, Chögyam '*The Myth of Freedom and The Way of Meditation*', Shambhala Publications Inc, Boston, 2005, p.70.

Watts, Alan, '*The Way of Zen*', Random House USA Inc, New York, 1957

Watts, Alan, '*The Wisdom of Insecurity: A Message for an Age of Anxiety*', Random House USA Inc, New York, 2011

Welwood, John, '*Towards a Psychology of Awakening*', Shambhala Publications Inc, Boston, 2002